T0321005

Phase Rule
and its applications

Phase Rule
and its applications

Suruchi

Assistant Professor
Department of Chemistry
Ramjas College, University of Delhi

Sheza Zaidi

Assistant Professor
Department of Chemistry
Ramjas College, University of Delhi

Manakin
PRESS

First published 2022
by CRC Press
4 Park Square, Milton Park, Abingdon, Oxon, OX14 4RN

and by CRC Press
6000 Broken Sound Parkway NW, Suite 300, Boca Raton, FL 33487-2742

© 2022 Manakin Press Pvt. Ltd.

CRC Press is an imprint of Informa UK Limited

British Library Cataloguing-in-Publication Data
A catalogue record for this book is available from the British Library

Library of Congress Cataloging-in-Publication Data
A catalog record has been requested

ISBN: 978-1-032-28306-7 (hbk)
ISBN: 978-1-003-29794-9 (ebk)

DOI: 10.1201/9781003297949

Manakin
PRESS

Brief Contents

Detailed Contents

Preface

Phase equilibrium is generally considered to be a difficult topic. We thought very hard about ways to make its study easier and this text is the result of our endeavors. This book tries to provide a complete overview of phase equilibrium at the level of Bachelor of Science degree in Chemistry, presenting the required definitions and mathematical concepts needed to understand the formalisms as well number of schematic phase diagrams. Theoretical descriptions of equilibrium conditions, the state of systems at equilibrium and the changes as equilibrium is reached, are all demonstrated graphically. This textbook deals with the theoretical basis of chemical equilibria and chemical changes and emphasizes the properties of phase diagrams. A large number of problems along with their solutions have been incorporated at the end, which provides a more easy approach towards understanding the concepts and fundamentals of the subject. This textbook is a valuable resource for advanced undergraduates and graduate students in chemistry

Efforts have been made to simplify every possible difficulty that might arise during construction of phase diagrams and cooling curves. The book will meet the basic requirement of the students. The subject matter has been presented in a logical and simple manner. Thus, we believe that the book will enable the students to gain confidence in the topic of phase equilibrium.

We welcome any suggestion or comment for further improvement in the subject matter covered in the book.

Authors

Acknowledgements

I, *Suruchi,* would like to thank my Nanaji, who has always been a constant source of inspiration and guided me to traverse the path of righteousness in my life.

I want to thank my husband, Manish for providing me unfailing support & continuous encouragement which helped in completion of this book. Special thanks goes to my daughter Saanvi for cooperating with me inspite of all the time it took me away from her.

I am grateful to my parents for their support and blessings.

Finally, I cannot forget the kind of help and encouragement I got from my most wonderful friend Oindrila De.

I, *Sheza*, would like to thank God who gave me the power to believe in my passion and pursue my dreams.

I am so thankful to my husband, Kausar, that I have you in my corner pushing me when I am ready to give up. A special thanks to my son, Faiz, you are the best thing that ever happened to me and made my life more meaningful than it ever was.

Most importantly, none of this would have been possible without the love and patience of my parents, who have been a constant source of love, concern, support and strength all these years. I would like to express my heart-felt gratitude to my sisters, Lubna and fauzia, who actually aided me and encouraged me throughout this endeavour.

I would like to offer my sincere thanks to my labmate and friend Dr. Suruchi, who gave me this wonderful opportunity and kept motivating me in my tough times.

We express our heartfelt gratitude to our guide, Prof. Rita Kakkar, who inculcated in us the spirit of scientific temperament.

We also express our thanks to all our colleagues at Ramjas College for comments and valuable suggestions towards the improvement of the book.

Lastly, we express our gratitude to the Manakin Press for their expertise in polishing the manuscript.

1

Phase Equilibria: The Phase Rule

1.1 INTRODUCTION

The phase rule was discovered by Josiah Williard Gibbs in 1876 is an important generalisation dealing with the behaviour of heterogeneous systems in equilibrium. It describes the degrees of freedom available to describe a particular system having different phases and substances. If pressure and temperature are the state variables, the rule can be written as:

$$F = C - P + 2$$

where, F = Number of independent variables (degrees of freedom)

C = Number of components and

P = Number of stable phases in the system.

This rule, if properly applied, has no exception.

Importance

Different types of system can be observed in daily life. For example, if a crystal of ice is in a glass of water and the same ice is in water closed in a bottle. The system now tries to acquire the equilibrium by changes in temperature, pressure and concentration. In the former case melting of ice depends on the temperature of water, the water in turn evaporates into vapour, but the glass allows the vapours to escape and hence there will be constant change in volume of water and the system will therefore never achieve an equilibrium state. However, in the latter case, since bottle is a closed system, it does not allow the vapours to escape. Therefore, melting of ice depends on temperature of water and hence volume remains constant. Thus, in order to define a particular system, to define the state of each phase in system we have to know about fixed number of variables. These variables can easily be predicted by using phase rule.

Homogeneous and Heterogeneous Equilibrium

If a system is uniform throughout in its chemical composition and physical state, it is called homogeneous.

e.g.: water, ice, water vapour, sugar dissolved in water, NaCl in water etc.

On the other hand, if a system is composed of more than one phase and have different chemical and physical properties and these phases are marked off and distinguished from each other by boundaries, such a system is said to be heterogeneous.

e.g.: — A cube of ice in water

— Ice, water and vapour in equilibrium. (Three phases, each in itself is homogeneous and can be mechanically separated from one other.)

1.2   TERMS USED IN PHASE RULE

1.2.1 Phase

A phase is defined as any homogeneous part of the system that is chemically and physically bounded by a distinct interface with other phases and physically separated from other phases.

Various Types of Phases:

(*i*) **Pure Substances**: A substance that has a fixed chemical composition throughout is a pure substance. Such as water, air and nitrogen (each made up of one chemical species and hence considered as one-phase system). Ice, water and vapour are three phases of the same chemical substance *i.e.*, water (different physical properties).

(*ii*) **Mixture of Gases**: All gases are miscible with one another in all proportions. Thus, a gaseous mixture is one phase system.

(*iii*) **Miscible Liquids**: Two completely miscible liquids forms a uniform solution. Therefore, a mixture of water and ethanol is a one-phase system.

(*iv*) **Non-Miscible Liquids**: A mixture of two non-miscible liquids forms two separate layers. Thus, a mixture of chloroform and water is a two-phase system.

(*v*) **Aqueous Solution**: The aqueous solution of a solid substances is uniform throughout and is therefore considered as one-phase system. However, saturated solution of NaCl in contact with excess of solid NaCl is not homogeneous and considered as two-phase system.

(*vi*) **Mixture of Solids**: A heterogeneous mixture of solid substances constitutes as many phases as there are substances present.

e.g.: (a) Decomposition of $CaCO_3$ (s)

$$CaCO_3 \text{ (s)} \rightleftharpoons CaO \text{ (s)} + CO_2 \text{ (g)}$$

Solid phases: $CaCO_3$ + CaO

Gaseous phase: CO_2

Total phases $\Rightarrow$ 2 + 1 = 3.

(b) Similarly, in the equilibrium reaction

$$Fe \text{ (s)} + H_2O \text{ (g)} \rightleftharpoons FeO \text{ (s)} + H_2 \text{ (g)}$$

There are two solid phases Fe and FeO and one gaseous phase consisting of H_2O (g) and H_2 (g). Thus, three phases exist in equilibrium.

(vii) Sulphur exists in two well-known crystalline forms-rhombic or octahedral, and monoclinic or prismatic sulphur. Both the forms have same chemical composition but different physical properties. Thus, a mixture of above two forms of sulphur is a two-phase system.

1.2.2 Components

The number of components of a system at equilibrium is defined as the least number of independent chemical constituents in terms of which the composition of every phase can be expressed by means of a chemical equation.

(i) **Ice-Water-Vapour System**:

$$Ice \text{ (s)} \rightleftharpoons Water \text{ (l)} \rightleftharpoons Vapour \text{ (g)}$$

The number of constituents taking part in the equilibrium is only one, viz. the chemical substance water. Hydrogen and oxygen, the constituents of water are not to be regarded as components, because they are not present in the system in a state of real equilibrium but they are combined in a definite proportion to form water, and their amounts, therefore, cannot be varied independently.

(ii) **Mixture of Gases**: The number of components in a mixture of gases is equal to the number of gases present in the mixture because all the gases are made up of different chemical substances.

e.g.: Mixture of O_2 and $N_2 \Rightarrow xO_2 + yN_2$

Two different chemical substances (O_2 and N_2) are present in a mixture of O_2 and N_2 and hence it is a two-phase system.

(iii) **The Sulphur System**: It consists of four phases rhombic, monoclinic, liquid and vapour, the chemical composition of all phases is S. Hence, it is one component system.

(iv) **Decomposition of $CaCO_3$ (s)**:

$$CaCO_3 \text{ (s)} \rightleftharpoons CaO \text{ (s)} + CO_2 \text{ (g)}$$

When equilibrium has been established, there are three different substances present $CaCO_3$, CaO and CO_2 and these are the constituents of the system between which equilibrium exist. These constituents take part in equilibrium, they are not regarded as components, as they are not mutually independent. On the contrary, the different phases are related to one another, and if two of these are taken, the composition of the third is defined by the equation:

$$CaCO_3 = CaO + CO_2$$

Of the three constituents, when the system is in equilibrium, only two are independently variable. Therefore, in order to express the composition of each phase, two of these constituents are required.

Thus, if two components $CaCO_3$ and CaO are taken, the composition of each phase can be expressed as:

$$CaCO_3 = CaCO_3 + OCaO$$
$$CaO = CaO + OCaCO_3$$
$$CO_2 = CaCO_3 - CaO$$

Similar expressions would be obtained if $CaCO_3$ and CO_2/CaO and CO_2 are taken as components.

So, two species out of three are sufficient to express the composition of all the three phases. Hence, it is a two-component system.

(v) **Dissociation of NH_4Cl:**

$$NH_4Cl\ (s) \rightleftharpoons NH_3(g) + HCl\ (g)$$

Available chemical constituents are three. Because of the equilibrium condition the number of independent components is reduced by one. And also, because the products formed form a single phase and are formed in equal amounts, the number of independent components are further reduced by one.

$$\therefore \qquad\qquad C = 1$$
or $$\qquad\qquad P = 2$$
$$F = 1 - 2 + 2 = 1$$

OR

The system consists of two phases solid NH_4Cl and gaseous mixture containing NH_3 and HCl. However, the constituents of the mixture are in the same proportion in which they are combined in solid NH_4Cl. The composition of both the phases therefore be expressed in terms of same chemical identity NH_4Cl. Thus, the dissociation of NH_4Cl is a one component system.

(vi) **Decomposition of PCl_5:**

$$PCl_5\ (s) \rightleftharpoons PCl_3\ (s) + Cl_2\ (g)$$

Because of the equilibrium condition, the number of independent components is reduced by one.

$$\therefore \qquad C = 2$$
$$P = 3$$
$$F = 2 - 3 + 2 = 1$$

(vii) Aqueous Solution of NaCl:

$NaCl + H_2O$: The system is composed of two constituents $i.e.$, NaCl and H_2O. Hence, it is a two-component system.

(viii) Dissociation of $CuSO_4 \cdot 5H_2O$ (s):

$$CuSO_4 \cdot 5H_2O \text{ (s)} \rightleftharpoons CuSO_4 \cdot 3H_2O \text{ (s)} + 2H_2O \text{ (g)}$$

The composition of each phase can be represented by the components, $CuSO_4$ and H_2O.

Hence, it is a two component system.

1.2.3 Degrees of Freedom or Variability of a System

By the term "degree of freedom" is meant to be the minimum number of variable factors, such as temperature, pressure and concentration of the component, which must be arbitrarily fixed in order that the condition of the system may be perfectly defined.

If, $\qquad F = 0 \rightarrow$ The system is nonvariant or invariant

$\qquad F = 1 \rightarrow$ The system is univariant or monovariant

$\qquad F = 2 \rightarrow$ The system is bivariant.

A knowledge of degrees of freedom is of essential importance in studying the condition and behaviour of a system and it is the great merit of the Phase Rule.

(i) **Water (l)** $\rightleftharpoons$ **Water Vapour (g)**: In this case, only temperature is sufficient to specify the state of the system since when temperature is fixed, vapour pressure of the system is automatically fixed. Hence, degree of freedom is one and the system is univariant.

(ii) **Ice (s)** $\rightleftharpoons$ **Water (l)** $\rightleftharpoons$ **Water Vapour (g)**: All the three phases are in equilibrium only at a particular temperature (co-exist at freezing point of water) and pressure, hence no condition need to be specified. The system is, therefore, non-variant or has no degrees of freedom.

(iii) For a system consisting of **water vapour phase** only, we need both the temperature and pressure to define the system completely. Hence, degree of freedom is two and the system is bivariant

(iv) For a system consisting of:

$$\textbf{NaCl (s)} \rightleftharpoons \textbf{NaCl (aq)} \rightleftharpoons \textbf{Water Vapour (g)}$$

We must state either the temperature or pressure, because the saturation solubility is fixed at a specific temperature or pressure. Hence, the system is univariant.

(*v*) For a **gaseous mixture of N$_2$ and H$_2$**, both temperature and pressure needs to be specified, because if pressure and temperature are fixed, the volume automatically becomes definite. Hence, for a gaseous system, two factors must be stated in order to define it completely and thus, it has two degrees of freedom (or bivariant).

1.3 ADVANTAGES OF PHASE RULE

1. It is applicable to both chemical and physical equilibrium.
2. It is applicable to macroscopic systems and hence no information is required regarding molecular micro structure.
3. We can conveniently classify equilibrium states in terms of phases, components and degrees of freedom.
4. The behaviour of system can be predicted under different conditions.
5. According to phase rule, different systems behave similarly if they have same degrees of freedom.
6. Phase rule helps in deciding under a given set of conditions:
   (*a*) Existence of equilibrium among various substances,
   (*b*) Inter convergence of substances,
   (*c*) Disappearance of some of the substances.

1.4 LIMITATIONS OF PHASE RULE

1. It is applicable only for the systems which are in equilibrium.
2. Only three degrees of freedom *viz.*, temperature, pressure and composition are allowed to influence the equilibrium systems.
3. Under the same conditions of temperature and pressure, all the phases of the system must be present.
4. It considers only the number of phases, rather than their amounts.

1.5 DERIVATION OF PHASE RULE

1.5.1 Phase Rule for a Non-Reactive system

Consider a heterogeneous system of P phases containing C components in equilibrium at constant temperature and pressure. (Fig. 1.1)

Number of variables to be known to define the system completely.

(*a*) Number of concentration variables to describe P phases = PC

(*b*) Temperature variable   = 1

(*c*) Pressure variable   = 1

      Total number of variables   = $PC + 2$

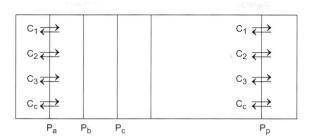

Fig. 1.1: *C* Components distributed in each of *P* phases.

The equations available are:

(*a*) Mole fractions in each phase must sum to unity.

$$x_1^{(a)} + x_2^{(a)} + x_3^{(a)} + \cdots x_c^{(a)} = 1$$

$$x_1^{(b)} + x_2^{(b)} + x_3^{(b)} + \cdots x_c^{(b)} = 1$$

$$\vdots$$

$$x_1^{(P)} + x_2^{(P)} + x_3^{(P)} + \cdots x_c^{(P)} = 1$$

Since, there is one such equation for each phase this gives *P* equations.

(*b*) The chemical potential of each component must be the same in every phase

$$\mu_1(a) = \mu_1(b) = \mu_1(c) = \ldots \mu_1(P)$$
$$\mu_2(a) = \mu_2(b) = \mu_2(c) = \ldots \mu_2(P)$$

$$\vdots$$

$$\mu_c(a) = \mu_c(b) = \mu_c(c) = \ldots \mu_c(P)$$

For each component there are $(P-1)$ such equations. Thus, for *C* components a total (equations) of $C\,(P-1)$ intensive variables are fixed by equilibrium conditions:

The number of degrees of freedom, is therefore.

$$F = [\text{Total number of variables}] - [\text{Equations available}]$$
$$= PC + 2 - [P + C\,(P-1)]$$
$$= PC + 2 - [P + PC - C]$$
$$= \cancel{PC} + 2 - P - \cancel{PC} + C$$
$$\boxed{F = C - P + 2}$$

⇒ This is the mathematical statement of the phase rule.

1.5.2 Phase Rule for a Reactive System

Consider a system of C constituents and P phases in equilibrium at constant temperature and pressure. Let us assume that four of the constituents are capable of undergoing the reaction:

$$v_1 A_1 + v_2 A_2 \rightleftharpoons v_3 A_3 + v_4 A_4$$

Total number of variables needed to define the system completely and the number of equations available at equilibrium.

Number of concentration variables to describe P phases $= PC$

Temperature, pressure variables $= 2$

Total number of variables $= PC + 2$

The equations available are:

(*a*) Equation relating the mole fraction for each phase

$$x_1^{(a)} + x_2^{(a)} + x_3^{(a)} + \cdots x_c^{(a)} = 1$$
$$x_1^{(b)} + x_2^{(b)} + x_3^{(b)} + \cdots x_c^{(b)} = 1$$
$$\vdots$$
$$x_1^{(P)} + x_2^{(P)} + x_3^{(P)} + \cdots + x_c^{(P)} = 1$$

There are P equations for P phases.

(*b*) Chemical Potential equations for each component

$$\mu_1(a) = \mu_1(b) = \mu_1(c) = \ldots = \mu_1(P)$$
$$\mu_2(a) = \mu_2(b) = \mu_2(c) = \ldots = \mu_2(P)$$

For C components in P phases, there are $C(P - 1)$ equations.

(*c*) *For a reactive system, one more condition needs to be satisfied. At equilibrium, the reaction potential $\widetilde{\Delta G}$ is zero.*

This gives $v_3\mu_3 + v_4\mu_4 - v_1\mu_1 - v_2\mu_2 = 0$

Thus, we get one more equation ($= 1$)

Total number of equations available $= P + C(P - 1) + 1$

$\therefore$ $F = $ Number of variables $-$ Equations available

$$= PC + 2 - [P + C(P - 1) + 1]$$
$$= (C - 1) - P + 2$$

If in a system, two independent reactions are possible, then

$$F = (C - 2) - P + 2$$

In general,

$$F = (C - r) - P + 2$$

$r = $ Number of independent reactions taking place in the system.

If, in general, a reaction system is having r independent reactions and z independent restricting conditions, then the number of equations that are available would be as follows:

Total number of equations

$$= \text{(Equations due to condition of mole fraction)}$$
$$+ \text{(Equation due to chemical potential)}$$
$$+ \text{(Equation due to independent chemical reactions)}$$
$$+ \text{(Equation due to restricting condition)}$$
$$= P + C(P-1) + r + z$$

Thus, $$F = (PC+2) - [P + C(P-1) + r + z]$$
$$= (C - r - z) - P + 2$$
$$F = C' - P + 2$$

where, $C' = C - r - z$ = number of components of the system. It is thus equal to the total number of constituents present in the system less than the number of independent chemical reactions and the number of independent restricting conditions.

1.5.3 Phase Rule when One of the Component is Missing from One Phase [One of the Component is present only in (P – 1) Phases]

Consider a system containing C components and P phases under equilibrium at constant temperature and pressure. Now, a component is missing from phase one and is present in only $P - 1$ phases. As one of the component is excluded from one phase, the number of concentration variables is decreased by one.

Number of concentration variables $\quad = CP - 1$

Pressure, temperature variables $\quad = 2$

$\therefore$ Total number of variables $\quad = CP + 1$

Number of possible equations is also decreased by one *i.e.*,

Number of phase equations $\quad = P$

Number of chemical potential equations for

$C - 1$ components in $P - 1$ phases $\quad = (C - 1)(P - 1)$

For one component in $P - 1$ phases $\quad = P - 2$

$\therefore$ Total number of equations $\quad = P + \{(C - 1)(P - 1) + (P - 2)\}$
$$= \{P + [C(P - 1) - 1]\}$$
$$= \{P + [CP - C - 1]\}$$

Degrees of freedom   =Total number of variables

                                        − Number of equations available

$$F = CP + 1 - \{P + [CP - C - 1]\}$$

$$= CP + 1 - \{P + CP - C - 1\}$$

$$= CP + 1 - P - CP + C + 1$$

$$= C - P + 2$$

i.e. phase rule remains the same whether each constituent is present in every phase or not. This means that phase rule is valid under all conditions of distribution provided that equilibrium exists in the system.

Examples to compute the number of components in a system.

Example (*i*) A solution containing Na^+, Cl^+, Ag^+, NO_3^-, AgCl (s) and H_2O.

$$NaCl \rightleftharpoons Na^+ + Cl^-$$

$$NaNO_3 \rightleftharpoons Na^+ + NO_3^-$$

$$AgNO_3 \rightleftharpoons Ag^+ + NO_3^-$$

$$AgCl \rightleftharpoons Ag^+ + Cl^-$$

$$H_2O \rightleftharpoons H^+ + OH^-$$

reactions = 5; constituents = 11

$$x_{Ag^+} + x_{Na^+} = x_{Cl^-} + x_{NO_3^-}$$

$$x_{H^+} = x_{OH^-}$$

$$\therefore \qquad z = 2$$

components (C) = constituents (C') − reactions (r) − electro neutrality condition (z)

$$= 11 - 5 - 2 = 4$$

$$\therefore \qquad C = 4$$

Example (*ii*) A solution containing H^+, OH^-, Na^+, Cl^-, Ag^+, NO_3^-, AgCl (s), H_2O

$$NaCl \rightleftharpoons Na^+ + Cl^-$$

$$NaNO_3 \rightleftharpoons Na^+ + NO_3^-$$

$$NaOH \rightleftharpoons Na^+ + OH^-$$

$$AgCl \rightleftharpoons Ag^+ + Cl^-$$

$$AgNO_3 \rightleftharpoons Ag^+ + NO_3^-$$

$$AgOH \rightleftharpoons Ag^+ + OH^-$$

$$HCl \rightleftharpoons H^+ + Cl^-$$

$$HNO_3 \rightleftharpoons H^+ + NO_3^-$$

$$H_2O \rightleftharpoons H^+ + OH^-$$

$$C' = 15, r = 9$$

$$x_{Na^+} + x_{Ag^+} + x_{H^+} = x_{Cl^-} + x_{NO_3^-} + x_{OH^-}$$

$$z = 1$$

$$C = C' - r - z$$

$$= 15 - 9 - 1 = 5$$

$$\boxed{C = 5}$$

Example (iii) A solution containing H_2O, Na^+, Cl^-, K^+, NO_3^-, NH_4^+, NH_3, H^+, OH^-

$$NaCl \rightleftharpoons Na^+ + Cl^-$$

$$NaNO_3 \rightleftharpoons Na^+ + NO_3^-$$

$$NaOH \rightleftharpoons Na^+ + OH^-$$

$$KCl \rightleftharpoons K^+ + Cl^-$$

$$KNO_3 \rightleftharpoons K^+ + NO_3^-$$

$$KOH \rightleftharpoons K^+ + OH^-$$

$$NH_4Cl \rightleftharpoons NH_4^+ + Cl^-$$

$$NH_4NO_3 \rightleftharpoons NH_4^+ + NO_3^-$$

$$NH_4OH \rightleftharpoons NH_4^+ + OH^-$$

$$H_2O \rightleftharpoons H^+ + OH^-$$

$$HCl \rightleftharpoons H^+ + Cl^-$$

$$HNO_3 \rightleftharpoons H^+ + NO_3^-$$

$$C' = 19, r = 12$$

$$x_{NH_4^+} + x_{K^+} + x_{Na^+} + x_{H^+} = x_{OH^-} + x_{NO_3^-} + x_{Cl^-}$$

$$z = 1$$

$$C = C' - r - z = 19 - 12 - 1 = 6$$

$$\boxed{C = 6}$$

Example (iv) An aq. solution containing H_3PO_4, $H_2PO_4^-$, HPO_4^{2-}, PO_4^{3-}, Na^+, H^+ at 1 atm. pressure

$$H_3PO_4 \rightleftharpoons 3H^+ + PO_4^{3-}$$

$$NaH_2PO_4 \rightleftharpoons Na^+ + H_2PO_4^-$$

$$H_2PO_4^- \rightleftharpoons H^+ + HPO_4^{2-}$$

$$HPO_4^{2-} \rightleftharpoons H^+ + PO_4^{3-}$$

$$H_2O \rightleftharpoons H^+ + OH^-$$

$$\left. \begin{array}{l} \\ \\ \\ \\ \\ \end{array} \right\} \quad \begin{array}{l} C' = 9 \\ r = 5 \end{array}$$

$$x_{\text{Na}^+} = x_{\text{PO}_4^{3-}} + x_{\text{HPO}_4^{2-}} + x_{\text{H}_2\text{PO}_4^-}$$

$$x_{\text{H}^+} = 2x_{\text{PO}_4^{3-}} + x_{\text{HPO}_4^{2-}} + x_{\text{OH}^-}$$

$\therefore$

$$z = 2$$

$$C = C' - r - z$$

$$= 9 - 5 - 2$$

$$= 2$$

❑❑❑

2

One Component Systems

2.1 INTRODUCTION

For one component system, the degrees of freedom is given by:

$$F = C - P + 2$$
$$= 1 - P + 2$$
$$= 3 - P$$

(*i*) If a one-component system has one phase only:

The value of F becomes:

$$F = 3 - P$$
$$= 3 - 1 = 2$$

that is, the system is bivariant. Therefore, two variables have to be specified *i.e.*, temperature and pressure to define the system completely.

(*ii*) If one-component system has two phases:

$$F = 3 - P$$
$$= 3 - 2 = 1$$

and anyone of the variable (temperature or pressure) is to be specified to define the system completely. If temperature is specified, then pressure will have a definite value and vice-versa.

(*iii*) If one-component system has three phases:

$$F = 3 - P$$
$$= 3 - 3 = 0$$

and system is invariant *i.e.,* three phases can exist in equilibrium at a definite value of temperature and pressure.

Thus, the maximum number of phases at equilibrium for one component system is three and the maximum degree of freedom is two. Hence, the phase diagram of one-component system can be drawn in two dimensions using temperature and pressure as variables.

2.2 CLAPEYRON EQUATION

Clapeyron discovered an important fundamental equation which helps in finding the extensive application in one component, two phase system. The Clapeyron equation is a way of characterizing a discontinuous phase transition between two phases of matter of a single constituent.

Consider any two phases of a same substance in equilibrium with each other at a given temperature and pressure. The two phases may be A and B. Let us suppose a pure substance which is initially in phase A changes to phase B. Both the phases are in equilibrium with each other at a given temperature and pressure.

Let G_A be the free energy per mole of the substances in initial phase A and G_B be the free energy per mole of the substances in phase B.

Since, the two phases are in equilibrium:

$$G_A = G_B$$

Hence, there will be no free energy change $i.e.$,

$$\Delta G = G_B - G_A = 0$$

If the temperature of the above system is raised from T to $T + dT$ as well as pressure from P to $P + dP$ so that equilibrium is maintained. Now the free energy per mole of the substance in phase A be $G_A + dG_A$ and that in phase B is $G_B + dG_B$. Since, the system is still in equilibrium then,

$$G_A + dG_A = G_B + dG_B \qquad \qquad ...(1)$$

Now, since $\qquad\qquad G_A = G_B$

$\therefore \qquad\qquad\qquad dG_A = dG_B \qquad\qquad\qquad\qquad ...(2)$

According to thermodynamics,

$$dG = -SdT + VdP \qquad\qquad ...(3)$$

Making use of equation (2) in equation (1)

$$dG_A = -S_A dT + V_A dP \qquad\qquad (4)$$

$$dG_B = -S_B dT + V_B dP \qquad\qquad ...(5)$$

(S_A and V_A are molar entropy and molar volume in phase A and S_B and V_B are in phase B.)

Therefore, $\quad -S_A dT + V_A dP = -S_B dT + V_B dP$

$$dT(S_B - S_A) = dP(V_B - V_A)$$

$$\boxed{\dfrac{dP}{dT} = \dfrac{S_B - S_A}{V_B - V_A} = \dfrac{\overline{\Delta S}\ \text{trans}_{(A\to B)}}{\overline{\Delta V}\ \text{trans}_{(A\to B)}}}$$

where, $\overline{\Delta S}$ trans and $\overline{\Delta V}$ trans are respective changes in entropy and volume of system when 1 mole of a substance is transformed from phase A to phase B.

This equation is known as Clapeyron equation.

2.3. APPLICATIONS OF CLAPEYRON EQUATION

1. Solid-Liquid Equilibrium: When 1 mole of a substance is transformed from solid phase to liquid phase, Clapeyron equation becomes:

$$\frac{dP}{dT} = \frac{\Delta \overline{S}_{fus}}{\Delta \overline{V}_{fus}} = \frac{\Delta \overline{H}_{fus}}{T.\Delta \overline{V}_{fus}} \qquad \boxed{\frac{dP}{dT} = \frac{S_B - S_A}{V_B - V_A} = \frac{\Delta \overline{S} \text{ trans}_{(A \to B)}}{\Delta \overline{V} \text{ trans}_{(A \to B)}}}$$

or

$$dP = \frac{\Delta \overline{H}_{fus}}{\Delta \overline{V}_{fus}} \frac{dT}{T}$$

Integrating within limits from P_1 to P_2 and from T_1 to T_2.

$$\int_{P_1}^{P_2} dp = \frac{\Delta \overline{H}_{fus}}{\Delta \overline{V}_{fus}} \int_{T_1}^{T_2} \frac{dT}{T}$$

Where T_1 and T_2 are the melting points at pressures P_1 and P_2. If $\Delta \overline{H}_{fus}$ and $\Delta \overline{V}_{fus}$ are considered independent of temperature and pressure, the above equation becomes:

$$P_2 - P_1 = \frac{\Delta \overline{H}_{fus}}{\Delta \overline{V}_{fus}} \ln \left(\frac{T_2}{T_1} \right)$$

Since the difference $T_2 - T_1$ is usually small, the logarithm term may be written as:

$$\ln \left(\frac{T_2}{T_1} \right) = \ln \left(\frac{T_1 + T_2 - T_1}{T_1} \right)$$

$$= \ln \left(1 + \frac{T_2 - T_1}{T_1} \right)$$

$$\simeq \frac{T_2 - T_1}{T_1}$$

$$= \frac{\Delta T}{T_1}$$

Thus,

$$\boxed{\Delta P = P_2 - P_1 = \frac{\Delta \overline{H}_{fus}}{\Delta \overline{V}_{fus}} . \frac{\Delta T}{T_1}}$$

2. Solid-Vapour Equilibrium: Taking the Clapeyron equation for solid-gas equilibrium:

$$\frac{dP}{dT} = \frac{\Delta \overline{S}_{sub}}{\Delta \overline{V}_{sub}} = \frac{\Delta \overline{H}_{sub}}{T \Delta \overline{V}_{sub}} \qquad \left[\Delta \overline{V}_{sub} = V_{m,g} - V_{m,s} \right]$$

where, $\Delta \overline{S}_{sub}$ = Entropy of sublimation per mole

and $\Delta \overline{V}_{sub}$ = Volume per mole when 1 mole of a substance is transformed from solid phase to vapour phase.

Now, making an approximation:

$$V_{m,g} - V_{m,s} \simeq V_{m,g}$$

Since

$$V_{m,g} \gg V_{m,s}$$

i.e. we can ignore the molar volume of the condensed phase compared to the gas.

Assuming the vapour phase to be ideal,

$$V_{m,g} = \frac{RT}{P}$$

$\Rightarrow$

$$\frac{dP}{dT} = \frac{\Delta \overline{H}_{sub}}{TV_{m,g}} = \frac{\Delta \overline{H}_{sub}}{T.\dfrac{RT}{P}}$$

$$\frac{dP}{dT} = \frac{\Delta \overline{H}_{sub}.P}{RT^2}$$

$$\frac{dP}{P} = \frac{\Delta \overline{H}_{sub}}{R} \frac{dT}{T^2}$$

$$\boxed{\frac{d \ln P}{dT} = \frac{\Delta \overline{H}_{sub}}{RT^2}}$$

This is the Clapeyron Equation for solid-vapour equilibrium.

We can make another approximation:

Assuming $\Delta \overline{H}_{sub}$ independent of T,

$$\int_{P_1}^{P_2} \frac{dP}{P} = \frac{\Delta \overline{H}_{sub}}{R} \int_{T_1}^{T_2} \frac{dT}{T^2}$$

$$ln \left(\frac{P_2}{P_1} \right) = \frac{-\Delta \overline{H}_{sub}}{R} \left[\frac{1}{T_2} - \frac{1}{T_1} \right]$$

$$\boxed{ln \left(\frac{P_2}{P_1} \right) = \frac{\Delta \overline{H}_{sub}}{R} \left[\frac{T_2 - T_1}{T_1 T_2} \right]}$$

This is the Integrated form of Clapeyron Equation.

3. Liquid-Vapour Equilibrium: For liquid-vapour equilibrium, replace $\Delta \overline{H}_{sub}$ with $\Delta \overline{H}_{vap}$

i.e. $\boxed{\dfrac{d \ln P}{dT} = \dfrac{\Delta \overline{H}_{vap}}{RT^2}}$ and $\boxed{ln \left(\dfrac{P_2}{P_1} \right) = \dfrac{\Delta \overline{H}_{vap}}{R} \left[\dfrac{T_2 - T_1}{T_1 T_2} \right]}$

The Clapeyron equation relates the temperature dependence of the vapour pressure of a liquid or a solid to $\Delta \overline{H}_{vap}$ or $\Delta \overline{H}_{sub}$ (respectively).

Numericals Based on Clapeyron Equation

Example 1: At 100°C, the specific volumes of water and steam are 1 cm^3 and 1670 cm^3 respectively. Calculate the change in vapour pressure of the system by 5°C change in temperature. The molar heat of vapourisation of water may be taken as 9.70 kcal.

Solution: Molar volume of liquid water, V_l

$$= 18 \text{ cm}^3 \text{ mol}^{-1}$$

$$= 18 \times 10^{-6} \text{ m}^3 \text{ mol}^{-1}$$

Molar volume of steam, V_g

$$= 18 \times 1670 \text{ cm}^3 \text{ mol}^{-1}$$

$$= 30060 \times 10^{-6} \text{ m}^3 \text{ mol}^{-1}$$

Heat of vapourisation, $\Delta \overline{H}_{vap}$

$$= 9{,}700 \text{ cal mol}^{-1} \times 4.184 \text{ J cal}^{-1}$$

$$= 40584.8 \text{ J mol}^{-1}$$

$$\frac{dP}{dT} = \frac{\Delta H_{vap}}{T(V_g - V_l)}$$

$$dT = 5 \text{ K}$$

$$T = 273 + 100$$

$$= 373 \text{ K}$$

$$\therefore \quad dP = \frac{40584.8 \text{ J mol}^{-1} \times 5 \text{ K}}{373 \text{ K} (30060 - 18) \times 10^{-6} \text{ m}^3 \text{ mol}^{-1}}$$

$$= \frac{202924 \text{ J}}{112056660 \times 10^{-6} \text{ m}^3}$$

$$= 0.018109 \times 10^6 \text{ Nm}^{-2}$$

$$= 0.1789 \text{ atm.}$$

$$dP = 135.9 \text{ mm Hg.}$$

Thus, vapour pressure of water increases by 135.9 mm Hg by 5°C rise in temperature at 100°C.

Example 2: The vapour pressure of water at 94.9°C is found to be 634 mm. What would be the vapour pressure at a temperature of 100°C. The heat of vapourisation in this range of temperature may be taken as 40593 J mol^{-1}.

Solution: $\quad \log\left(\dfrac{P_2}{P_1}\right) = \dfrac{\Delta \overline{H}_{vap}}{2.303 \, R}\left[\dfrac{T_2 - T_1}{T_1 T_2}\right]$

$$T_1 = 273 + 94.9 = 367.9 \text{ K}, P_1 = 634 \text{ mm}$$

$$T_2 = 273 + 100 = 373 \text{ K}, P_2 = ?$$

$$\Delta \overline{H}_{vap} = 40593 \text{ J mol}^{-1}$$

$$\log\left(\frac{P_2}{634 \text{ mm}}\right) = \frac{40593 \text{ J mol}^{-1}}{8.314 \text{ J} \times 2.303 \text{ K}^{-1} \text{ mol}^{-1}}\left[\frac{5.1 \text{ K}}{367.9 \text{ K} \times 373 \text{ K}}\right]$$

$$\boxed{P_2 = 759.7 \text{ mm Hg}}$$

Example 3: Suppose solid liquid equilibrium mixture is maintained at triple point temperature (0.0075°C) and triple point pressure (4.6 mm Hg). Calculate the change in temperature that should be made so that equilibrium is maintained at an external pressure of 1 atm.

Given: ρ (ice) = 0.917 g cm^{-3}

                   ρ (liquid) = 0.9998 g cm^{-3}

and $\Delta \overline{H}_{fus}$ = 6008.5 J mol^{-1}

Solution: $\dfrac{dP}{dT} = \dfrac{\Delta \overline{H}_{fus}}{T(V_{m, l} - V_{m, s})}$

$$(V_{m, l} - V_{m, s}) = \frac{18 \text{ g mol}^{-1}}{0.9998 \text{ g em}^{-3}} - \frac{18 \text{ g mol}^{-1}}{0.917 \text{ g em}^{-3}}$$

$$= -1.6256 \text{ cm}^3 \text{ mol}^{-1}$$

$$= -1.6256 \times 10^{-6} \text{ m}^3 \text{ mol}^{-1}$$

Therefore, $\dfrac{dP}{dT} = \dfrac{6008.5 \text{ J mol}^{-1}}{(273.0075 \text{ K})(-1.6256 \times 10^{-6} \text{ m}^3 \text{ mol}^{-1})}$

$$= -1.3538 \times 10^7 \text{ Pa K}^{-1}$$

or $dT = \dfrac{dP}{1.3538 \times 10^7 \text{ Pa K}^{-1}}$

$$dP = 760 \text{ mm Hg} - 4.6 \text{ mm Hg}$$

$$= 755.4 \text{ mm Hg}$$

$$= \frac{755.4}{760} \times 101.325 \times 10^3 \text{ Pa}$$

$$= 1.0071 \times 10^5 \text{ Pa}$$

$$dT = \frac{1.0071 \times 10^5 \text{ Pa}}{-1.3538 \times 10^7 \text{ Pa K}^{-1}}$$

$$dT = -0.00742 \text{ K}.$$

Example 4: The specific volumes of ice and water at 0°C are 1.0906 cm^3 and 1.0001 cm^3 respectively. Calculate the change in melting point of ice per atm increase of pressure? ΔH_{fus} of ice = 79.8 cal g^{-1}.

Solution:
$$V_{m, s} = 18 \times 1.0906 \text{ cm}^3$$
$$= 18 \times 1.0906 \times 10^{-6} \text{ m}^3$$
$$V_{m, l} = 18 \times 1.0001 \text{ cm}^3$$
$$= 18 \times 1.0001 \times 10^{-6} \text{ m}^3$$
$$T = 273 \text{ K}$$
$$\Delta \overline{H}_{fus} = 18 \text{ g mol}^{-1} \times 79.8 \text{ cal g}^{-1} \times 4.184 \text{ J cal}^{-1}$$
$$= 6009.9 \text{ J mol}^{-1}$$
$$dP = 1 \text{ atm.} = 101325 \text{ Nm}^{-2}$$
$$dT = ?$$
$$\frac{dT}{dP} = \frac{T(V_{m, l} - V_{m, s})}{\Delta H_{fus}}$$
$$= \frac{273 \text{ K}(-0.0906 \times 18 \times 10^{-6} \text{ m}^3)}{6009.9 \text{ J mol}^{-1}}$$
$$= \frac{273 \text{ K} \times 18 \times 10^{-6} \times (-0.0906) \text{m}^3 \times 101325 \text{ Nm}^{-2}}{6009.9 \text{ J mol}^{-1}}$$
$$\frac{dT}{dP} = -0.0075 \text{ K}$$

i.e., melting point of ice decreases with increase of pressure.

2.4   CONSTRUCTION OF PHASE DIAGRAM OF ONE COMPONENT SYSTEMS

Phase diagram is a graphical representation of the physical states of a substance under the different conditions of temperature and pressure. A typical phase diagram has pressure on the *y*-axis and temperature on the *x*-axis. As we cross the lines or curves on the phase diagram, a phase change occurs.

That is, at a point on line, it is possible for two (or three phases) to co-exist at equilibrium. In other regions of the plot, only one phase exists at equilibrium. A phase diagram is a common way to represent the various phases of a substance and the conditions under which each phase exists.

The figure shown below displays a typical phase diagram for one-component system (*i.e.*, one consisting of a single pure substance), the curves having been obtained from measurements at various pressures and temperatures.

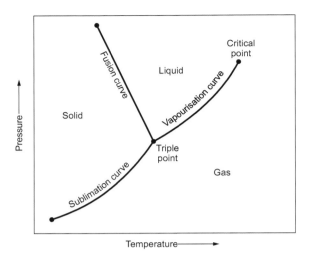

Fig. 2.1: Phase Diagram of One-component system

At any point in the areas separated by the curves, the pressure and temperature allow only one phase (solid, liquid, or gas) to exist. At any point on the curves, the temperature and pressure allow two phases to exist in equilibrium; solid and liquid, solid and vapour, or liquid and vapour. Along the line between solid and liquid, the melting temperatures for different pressures can be found. The junction of three curves is called the triple point. Phase diagrams are specific for each substance and mixture.

2.5 PHASE DIAGRAM OF ONE COMPONENT SYSTEMS

2.5.1 Phase Diagram of Water

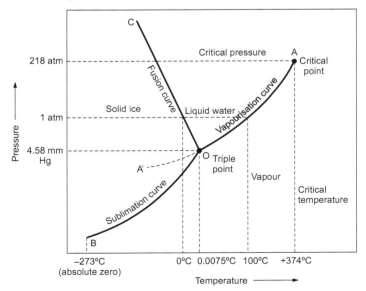

Fig. 2.2: Phase Diagram of Water

- Water exist in three possible phases: solid, liquid and vapour. Hence, there can be three forms of equilibrium,

1. Liquid $\rightleftharpoons$ Vapour

2. Solid $\rightleftharpoons$ Vapour   Each equilibria involves two phases.

3. Solid $\rightleftharpoons$ Liquid

- The phase diagram of water is divided into three **areas**: AOB, AOC and BOC.

Areas	Representation	No. of Phases
AOC	Condition of water existence	1
AOB	Condition of water vapour existence	1
BOC	Condition of ice existence	1

In all the areas, system has one phase and one component. Therefore, $P = 1$, $C = 1$ and $F = C - P + 2 = 1 - 1 + 2 = 2$. Hence the system is bivariant.

- The phase diagram has three **lines** or **curves**:

1. Equilibrium between Liquid and Vapour, Vaporisation Curve (OA): According to the phase rule, a system consisting of one component in two phases has one degree of freedom *i.e.*, univariant. Therefore, it will be possible for liquid water to co-exist with water vapour at different values of temperature and pressure, but if one of the variable factor *i.e.*, pressure, temperature or volume is fixed, the state of the system will then be defined. For any given temperature there exists only one vapour pressure. Similarly, for each vapour pressure, only one temperature can be maintained.

If the points indicating the values of pressure corresponding to different temperatures are joined, curve OA is obtained which represents the variation of pressure with temperature. This is called vapourisation curve of water.

Along this curve liquid water and vapour are in equilibrium *i.e.*, there are two phases

$\therefore$ $F = 1 - 2 + 2 = 1$

Hence, the system along curve OA is univariant.

The slope of the curve OA at any point is given by the Clapeyron equation:

$$\left(\frac{dp}{dT} \right)_{L \rightleftharpoons V} = \frac{\Delta \overline{H}_{vap}}{T\left(V_g - V_l\right)}$$

The curve OA has an upper limit at the critical pressure and critical temperature.

Since, with increase of pressure the density of vapour must increase, and since with rise of temperature the density of liquid must decrease, a point will be reached at which the density of liquid and vapour become identical, the system ceases to be heterogeneous and passes into one homogeneous phase. The temperature at which this occurs is called the *critical temperature*. To this temperature, there will correspond a certain definite pressure called *critical pressure*. The curve representing equilibrium between liquid and vapour must, therefore, end abruptly at critical point.

At temperature above this point no pressure, however great, can cause the formation of liquid phase.

$$\text{Critical Pressure} = 218 \text{ atm.}$$

$$\text{Critical Temperature} = 374°C$$

Above the critical temperature, it is impossible to condense a gas into a liquid just by increasing the pressure. The particles have too much energy for the intermolecular attractions to hold them together as a liquid.

(**Critical point**: A critical point is the end point of a phase equilibrium curve. The most prominent example is the liquid-vapour critical point, the end point of the pressure-temperature curve that designates conditions under which a liquid and its vapour can co-exist. At the critical point, phase boundaries vanish.)

2. Sublimation Curve of Ice (OB): The condition of equilibrium between ice and water vapour will be represented by a line or curve showing the change of pressure with temperature. Such a curve representing the condition of equilibrium between a solid and its vapour is called a sublimation curve, and is represented by *OB*.

At temperatures represented by any point on this curve, the solid ice will sublime into vapour without previously fusing. Its lower end B extends to absolute zero. Along this curve, ice and vapour are in equilibrium *i.e.*, two phases in equilibrium.

Therefore, $F = 1 - 2 + 2 = 1$ (univariant)

The slope of the curve at any point as given by Clapeyron equation:

$$\frac{dP}{dT} = \frac{\Delta \overline{H}_{sub}}{T(V_g - V_s)}$$

or
$$\ln \frac{P_2}{P_1} = \frac{\Delta \overline{H}_s}{R} \left\{ \frac{T_2 - T_1}{T_1 T_2} \right\} \qquad \text{[integrated form]}$$

3. Fusion Curve (OC): The curve representing the temperature and pressure at which ice and water are in equilibrium will represent the change of melting point with pressure. Such a curve is called the fusion curve represented by *OC*. It can be seen that as pressure is increased, melting point of ice is lowered. That is why the curve *OC* is inclined towards the pressure axis.

The slope of this line can be given by Clapeyron equation as:

$$\left(\frac{dP}{dT} \right)_{S \rightleftharpoons L} = \frac{\Delta \overline{H}_{fus}}{T(V_l - V_s)}$$

Since, density of ice is less than that of water, V_s is greater than V_l the expression on the right hand side is negative. Hence, dP/dT have a negative sign and hence the line *OC* is slightly tilted towards pressure axis.

Along this curve liquid water and ice are in equilibrium *i.e.*, two phases are in equilibrium.

Therefore, according to the phase rule,

$$F = 1 - 2 + 2 = 1, \text{ the system is univariant.}$$

[The slope of the fusion curve in case of water is negative. As a result, water can melt at temperature near its freezing point when subjected to pressure. The ease with which ice skaters glide across a frozen pond can be explained by the fact that the pressure exerted by their skates melts a small portion of the ice that lies beneath the blades.]

• **Equilibrium between ice, water and vapour (Triple Point)**: At a temperature of 0.0075°C and under a pressure of 4.58 mm Hg, three two curves (OA, OB and OC) intersect. At this point, therefore, ice, water and vapour are in equilibrium and the point is called a *triple point* represented by point O in phase diagram of water. At point O, there are three phases in equilibrium.

Therefore, $F = 1 - 3 + 2 = 0$ (invariant)

At the triple point, the slope of $S \rightleftharpoons V$ (curve OB) is greater than that of $L \rightleftharpoons V$ (Curve OA). This can be shown using Clapeyron equation:

$$\left(\frac{dP}{dT} \right)_{S \rightleftharpoons V} = \frac{\Delta \overline{H}_{sub}}{T(V_g - V_s)}$$

$$\left(\frac{dP}{dT} \right)_{L \rightleftharpoons V} = \frac{\Delta \overline{H}_{vap}}{T(V_g - V_l)}$$

$$V_g - V_s \rightleftharpoons V_g - V_l$$

At the triple point,

$$\Delta \overline{H}_{sub} = \Delta \overline{H}_{fus} + \Delta \overline{H}_{vap}$$

Since, $\Delta \overline{H}_{sub} > \Delta \overline{H}_{vap}$

$$\left(\frac{dP}{dT} \right)_{S \rightleftharpoons V} > \left(\frac{dP}{dT} \right)_{L \rightleftharpoons V}$$

• **Supercooled water, metastable state**: Sometimes, it is possible to cool water below its freezing point without the separation of ice. The water is then said to be supercooled and can be kept as such for long time if the presence of any solid particle is avoided. The system represented by point on the curve AO *i.e.*, liquid water in equilibrium with vapour, is cooled rapidly, ice may fail to form at the triple point and the vapour pressure of the liquid may continue along OA'. The liquid water $\rightleftharpoons$ vapour along the curve OA' is said to be in metastable equilibrium because as soon as a small particle of ice is brought in contact with supercooled liquid, entire liquid solidifies. It will be seen from the phase diagram that the curve

OA' lies above the curve *OB*. Thus, the metastable system has a higher vapour pressure than the stable one at the same temperature.

2.5.2 Phase Diagram of Carbon Dioxide

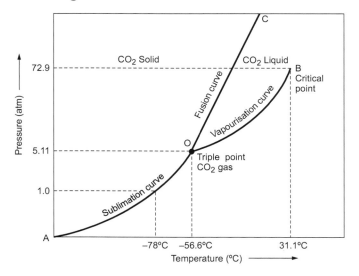

Fig. 2.3: Phase Diagram of Carbon Dioxide

- The diagram is divided into three **areas:**

Areas	Representation	No. of Phases
AOC	Condition of solid CO_2	1
AOB	Condition of CO_2 gas	1
BOC	Condition of liquid CO_2	1

In all the three areas, system has one phase and one component. Therefore, $P = 1$, $C = 1$ and $F = C - P + 2 = 1 - 1 + 2 = 2$.

- The phase diagram of CO_2 has three distinct **curves:**

1. Curve *OA*: Curve *OA* is the sublimation curve along which solid carbon dioxide is in equilibrium with the gas. Hence, $P = 2$, $C = 1$ *i.e.* $F = 1 - 2 + 2 = 1$.

2. Curve *OB*: Curve *OB* is the vaporisation curve along which liquid carbon dioxide is in equilibrium with gas. Therefore, $P = 2$, $C = 1$ and $F = 1$ *i.e.*, system along this curve is monovariant.

3. Curve *OC*: Curve *OC* is the fusion curve along which solid carbon dioxide is in equilibrium with liquid. Therefore, $P = 2$, $C = 1$ and $F = 1$ *i.e.*, system is monovariant.

In case of CO_2, the fusion curve (curve *OC*) slopes away from the pressure axis unlike water system which shows that with increase of pressure melting point of solid carbon dioxide increases.

According to clapeyron equation:

$$\left(\frac{dP}{dT}\right)_{S \rightleftharpoons L} = \frac{\Delta \overline{H}_{fus}}{T(V_l - V_s)}$$

As $V_l > V_s$ and $V_l - V_s$ is small, OC has a large positive slope.

• **Triple point O:** At point O, all the three phases of carbon dioxide are in equilibrium with one another. This point is the triple point and it occurs at $-56.6°C$ and 5.11 atm. At this point $P = 3$, $C = 1$ and $F = 1 - 3 + 2 = 0$ *i.e.*, system is invariant.

The triple point lies above atmospheric pressure, therefore it is not possible to get liquid carbon dioxide less than a pressure of 5.11 atm. At 1 atm, the liquid phase is not stable, the solid simply sublimates at $-78°C$. Thus, solid carbon dioxide is also known as dry ice.

The critical temperature for carbon dioxide is $31.1°C$ and the critical pressure is 72.9 atm. Above the critical temperature, the fluid is called super critical fluid.

2.5.3 Phase Diagram of Sulphur

Sulphur exists in two cystalline forms, rhombic (S_R) and monoclinic (S_M). Rhombic sulphur melts at a temperature of $114.5°C$ and monoclinic sulphur melts at $120°C$. At a temperature of $95.6°C$, the two forms can be transformed into one another. This temperature is known as the *transition temperature*.

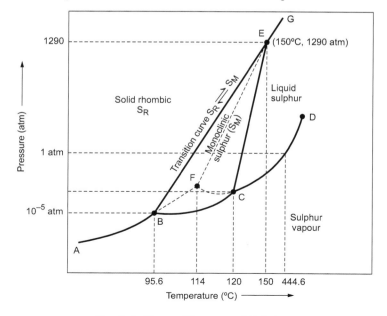

Fig. 2.4: Phase Diagram of Sulphur

- Sulphur exists in four possible phases:

1. Rhombic sulphur (S_R) $\left.\begin{array}{c}\\\\\\\\\end{array}\right\}$ Solid polymeric phases, but exists in different phases as they differ in their physical properties.

2. Monoclinic sulphur (S_M)

3. Sulphur liquid (S_L)

4. Sulphur vapours (S_V)

All the four phases are represented by one chemical identity *i.e.*, sulphur. Therefore, it is four phase and one component system.

- The phase diagram is divided into four **areas**:

Areas	Representation	No. of Phases
ABEG	Rhombic sulphur	1
BCEB	Monoclinic sulphur	1
DCEG	Sulphur liquid	1
ABCD	Sulphur vapour	1

In all the areas, system has one phase and one component. Therefore, $P = 1$, $C = 1$ and $F = 1 - 1 + 2 = 2$ *i.e.*, system is bivariant.

The phase diagram has the following **curves**:

1. Curve *AB*: Sublimation curve of rhombic sulphur. Along this curve rhombic sulphur and sulphur vapour are in equilibrium. System is monovariant.

$$S_R \rightleftharpoons S_V$$

2. Curve *BC*: Sublimation curve of monoclinic sulphur. Along this curve monoclinic sulphur and sulphur vapour are in equilibrium. It shows the variation of vapour pressure of monoclinic sulphur with temperature.

$$S_M \rightleftharpoons S_V$$

i.e., there are two phases and one component. Therefore, system is monovariant.

3. Curve *CD*: Vapour pressure curve of liquid sulphur. Along this curve liquid sulphur and its vapour exists in equilibrium, $S_L \rightleftharpoons S_V$. Again there are two phases and system is monovariant. It shows the variation of vapour pressure of liquid sulphur with temperature. The 1 atm line meets the curve *CD* at 444.6°C which is the boiling point of sulphur.

4. Curve *CE*: Fusion curve for monoclinic sulphur. Along the curve *CE*, the equilibrium between monoclinic and liquid sulphur exists and the system is monovariant. This curve gives the effect of pressure on the melting point of monoclinic sulphur. The melting or fusion of monoclinic sulphur is accompained by a slight increase in volume and therefore, melting point increases slightly with increase of pressure according to Clapeyron equation. Therefore, curve *CE* slopes away from the pressure axis.

5. Curve *BE*: Transition curve which gives the variation of pressure on the transition temperature of rhombic sulphur into monoclinic sulphur. Along the curve, rhombic sulphur and monoclinic sulphur are in equilibrium and system is monovariant.

$$\left(\frac{dP}{dT}\right)_{S_R \rightleftharpoons S_M} = \frac{\Delta H_t}{T(V_M - V_R)} \qquad [\Delta H_t = \text{molar heat of transition}]$$

Since, density of S_M is less than that of S_R

$$\therefore \qquad V_M > V_R$$

$$\therefore \qquad \left(\frac{dP}{dT}\right)_{S_R \rightleftharpoons S_M} = \text{positive}$$

Since, $\frac{dP}{dT}$ have a positive value and therefore the curve *BE* slopes away from the pressure axis.

6. Curve *EG*: Fusion curve of rhombic sulphur. Along this curve the two phases rhombic sulphur and liquid sulphur are in equilibrium and the system is monovariant.

• *Triple Points*

The phase diagram of sulphur has three triple points *viz*, *B*, *C* and *E*.

1. Triple point *B*: The three curves *AB*, *BC* and *BE* meet at point *B* where three phases, two solids and the vapour exist in equilibrium. *i.e.*, $S_R \rightleftharpoons S_M \rightleftharpoons S_V$. At point *B*, rhombic sulphur changes into monoclinic sulphur; therefore, temperature corresponding to point *B* is the transition temperature (95.6°C).

2. Triple point *C*: The three curves *BC*, *CE* and *CD* meet at point *C* at which three phases, *viz.*, monoclinic sulphur, liquid sulphur and sulphur vapour co-exist in equilibrium. *i.e.*, $S_M \rightleftharpoons S_L \rightleftharpoons S_V$. The temperature corresponding to point *C* is the melting point of monoclinic sulphur (120°C).

3. Triple point *E*: The three curves *BE*, *CE* and *EG* meet at point *E* where rhombic, monoclinic and liquid sulphur co-exist in equilibrium *i.e.*, $S_R \rightleftharpoons S_M \rightleftharpoons S_L$.

At any triple point, phase rule becomes:

$$F = 3 - P$$

$$= 3 - 3 = 0 \text{ (nonvariant)}.$$

• *Metastable Equilibrium*

Since, the conversion of rhombic sulphur into monoclinic sulphur is naturally slow. Therefore, if enough time for transformation is not allowed, there is a possibility that rhombic sulphur may pass well above the transition point without getting changed into monoclinic sulphur. In that case there are three phases and one component and diagram looks like that of water. Therefore, the first solid *i.e.*, rhombic sulphur will exist in metastable equilibrium with its vapour. Such a metastable equilibrium is represented by dotted lines.

Curve *BF* (Metastable sublimation curve of S_R): If a system on line *AB* is heated rapidly, it may happen that monoclinic sulphur does not appear at point *B* but the *V.P.* of the system continues along the line *BF*, which now represents metastable equilibrium between S_R and S_V. As there are two phases, therefore, $F = 1 - 2 + 2 = 1$ and hence system is monovariant.

Curve *CF* (Metastable vaporisation curve of supercooled S_L): If a system on *CD* is cooled rapidly, it may happen that monoclinic sulphur does not appear at point *C* and the system continues along line *CF* which represents metastable equilibrium between S_L and S_V *i.e.*, two phases and, therefore, system is monovariant.

Curve *EF* (Metastable fusion curve of S_R): If a system of rhombic sulphur at some high pressure is heated rapidly, transition to monoclinic sulphur might not occur on the line *BE* but the system might pass over directly to liquid phase. The system along curve *EF* represents metastable equilibrium $S_R \rightleftharpoons S_L$.

Again there are two phases and system is monovariant.

Metastable triple point *F*: The three metastable phases S_R, S_L and S_V co-exist in equilibrium at triple point *F*. At point *F*, the system is nonvariant.

$$F = 1 - 3 + 2 = 0$$

3

Two Component Systems

3.1 INTRODUCTION

When a single phase is present in a two-component system, $F = 3$.

$$F = 2 - 1 + 2 = 3$$

i.e., three variables must be specified to describe the system completely. Thus, in addition to the temperature and pressure, the concentration of one of the components has also to be given. For graphical representation of these variables, three co-ordinate axes at right angles to each other would be required. Therefore, phase diagram obtained would be a solid model (3-*D*).

For the sake of having simple phase diagram we generally consider only two variables, the third one being a constant *e.g.*, for a solid-liquid equilibrium, the gas phase is usually absent and the effect of pressure on equilibrium is very small. Thus, when a two component system consists of solid and liquid phases only, the effect of pressure may be disregarded. The system is then condensed system and the experimental measurements of temperature and concentration are usually carried out at 1 atm pressure. Since, degrees of freedom is reduced by one, phase rule may be written as:

$$F' = C - P + 1$$

This is the *reduced phase rule* for two component systems. The phase diagrams for such systems consists of temperature-concentration graphs since only temperature and concentration variables are taken into account.

3.2 TYPES OF PHASE DIAGRAMS IN TWO COMPONENT SYSTEMS

1. **Type *A***: In the liquid phase, the two components are completely miscible.

   1. Only the pure components crystallise from the solution.

2. The pure components crystallise from the solution and one of the solid exists in more than one crystalline form.

3. A solid compound stable upto its melting point is formed by two constituents.

4. A solid compound which decomposes before it reaches its melting point is formed by the two constituents.

5. The two components are completely miscible in the solid phase and form a series of solid solutions.

6. In the solid state, the two constituents are partially miscible and they form stable solid solutions.

7. Solid solutions formed by two constituents and are stable only upto a transition temperature.

2. Type *B*:

The two components are partially miscible in the liquid phase and only pure components crystallize from the solution.

3. Type *C*: Phase Diagrams of Aqueous solution of salts. An aqueous solution of a salt is an example of a two component system and we may have:

1. Formation of simple eutectic or cryohydrate.

2. Formation of compounds (hydrates) with congruent melting points.

3. Formation of compounds (hydrates) with incongruent points.

3.3   THERMAL ANALYSIS

The phase diagram showing different solid-liquid equilibrium systems can be drawn using thermal analysis or cooling curves. If we start with a liquid phase and allow it to cool in steady surroundings, the graph line that is obtained is a cooling curve and is plotted between **Temperature** versus **Time** for various compositions of a system.

A system of known composition is prepared, heated to get a melt, allowed to cool on its own and then its temperature is noted at regular intervals of time. Cooling curve is then plotted between temperature and time. Information regarding the initial and final solidification temperature is obtained from the breaks and halts in the cooling curves. The same procedure is repeated with different composition of the system. Thermal analysis method is applicable under all temperature conditions but is especially suitable for investigations at temperatures quite above and below room temperature.

3.3.1 Cooling Curve of a Pure Component

Pure component may be taken as A or (B) in liquid phase. The cooling of liquid takes place along ab and this cooling is smooth till solidification starts at b, where there are two phases and system is invariant $(F = C - P + 1; 1 - 2 + 1 = 0)$. Temperature remains constant along bd till entire liquid solidifies. This is due to the fact that rate of liberation of heat during solidification is equal to the rate of transfer of heat from system to surroundings. Along de cooling of solid takes place. System remains univariant along ab and de. $(F = C - P + 1; P = 1; 1 - 1 + 1 = 1)$.

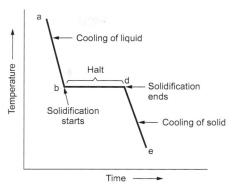

Fig. 3.1: Cooling curve of a pure component

3.3.2 Cooling Curve of a Mixture

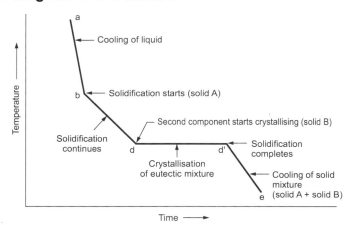

Fig. 3.2: Cooling curve of a Mixture

$$ab; F = 2 \qquad d'e; F = 1$$
$$b; F = 1$$
$$bd; F = 1$$
$$d; F = 0$$
$$dd'; F = 0$$
$$ab: (C - P + 1 = 2 - 1 + 1 = 2)$$
$$b: (C - P + 1 = 2 - 2 + 1 = 1)$$
$$bd: (C - P + 1 = 2 - 2 + 1 = 1)$$
$$d: (C - P + 1 = 2 - 3 + 1 = 0)$$

$$dd': (C - P + 1 = 2 - 3 + 1 = 0)$$
$$d': (C - P + 1 = 2 - 3 + 1 = 0)$$
$$d'e: (C - P + 1 = 2 - 2 + 1 = 1)$$

Cooling of liquid takes place along ab and at b solid A starts solidifying and the system becomes univariant. The rate of cooling along bd slows down due to liberation of heat during solidification and the cooling curve shows a distinct break at point b. The break point indicates the temperature at which first crystal of A is obtained. On further cooling, solution becomes saturated with respect to B and hence B also starts solidifying alongwith A and solution becomes invariant at d. The temperature remains constant along dd' until the solidification has taken place. Further cooling results in fall of temperature along $d'e$ where cooling of solid mixture takes place.

Thus, there are three breaks in the cooling curve. The first one occurs at freezing point of the mixture where the solid first commences to form. The second break occurs at eutectic point and third one when mixture gets completely solidified.

If we assume that at point d' the liquid is present in minute traces then F=0 (A $\rightleftharpoons$ B $\rightleftharpoons$ liquid) however, if we assume that at point d' the liquid has completely disappeared then in that case F=1 (A $\rightleftharpoons$ B, so the number of phases will be 2 instead of 3).

3.4   CONSTRUCTION OF PHASE DIAGRAM OF TWO COMPONENT SYSTEMS

Several mixtures of known composition are prepared. Equal amount of each mixture is then taken in the boiling tube and heated to melt. The mixture is then allowed to cool and temperature is noted with regular intervals of time. The cooling curves are drawn by plotting temperature (y-axis) against time (x-axis). A graph is then plotted between temperature (first break where solidification starts) and composition. A phase diagram of simple eutectic system is represented as:

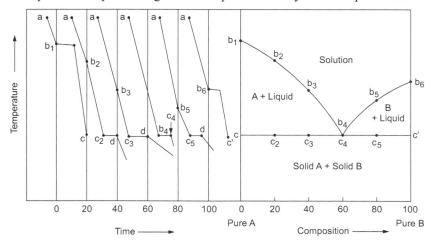

Fig. 3.3: Cooling curves and phase diagram constructed by thermal analysis.

3.4.1 Lever Rule

Lever rule is used to determine the composition of phases present in equilibrium and the relative proportion of phases to each other in binary systems.

Suppose there are two phases as a and b and the two components as 1 and 2, in equilibrium:

$$n_T = \text{Total number of moles present in system}$$

$$n_1^a = \text{Moles of component 1 in } a\text{-phase}$$

$$n_1^b = \text{Moles of component 1 in } b\text{-phase}$$

Similarly, $\quad n_2^a = \text{Moles of component 2 in } a\text{-phase}$

$$n_2^b = \text{Moles of component 2 in } b\text{-phase}$$

$$n_T = (n_1^a + n_1^b + n_2^a + n_2^b)$$

$$x_1 = \text{Mole fraction of component 1 in both phases.}$$

But, $\qquad x_1^a = \dfrac{n_1^a}{n_1^a + n_2^a}$

$$x_1^b = \dfrac{n_1^b}{n_1^b + n_2^b}$$

i.e. $\qquad n_1^a = x_1^a (n_1^a + n_2^a)$

$$n_1^b = x_1^b (n_1^b + n_2^b)$$

$$x_1 = \dfrac{n_1^a + n_1^b}{n_1^a + n_1^b + n_2^a + n_2^b}$$

$$x_1 = \dfrac{x_1^a (n_1^a + n_2^a) + x_1^b (n_1^b + n_2^b)}{n_1^a + n_1^b + n_2^a + n_2^b}$$

$$x_1(n_1^a + n_2^a) + x_1(n_1^b + n_2^b) = x_1^a (n_1^a + n_2^a) + x_1^b (n_1^b + n_2^b)$$

$$(n_1^a + n_2^a)(x_1 - x_1^a) = (n_1^b + n_2^b)(x_1^b - x_1)$$

Therefore; $\qquad \dfrac{(n_1 + n_2)^a}{(n_1 + n_2)^b} = \dfrac{x_1^b - x_1}{x_1 - x_1^a}$

i.e. $\qquad \dfrac{\text{No. of moles in phase } a}{\text{No. of moles in phase } b} = \dfrac{x_1^b - x_1}{x_1 - x_1^a}$

In order to determine the relative amounts of liquid and vapour phases with reference to the following phase diagram, lever rule can be applied as:

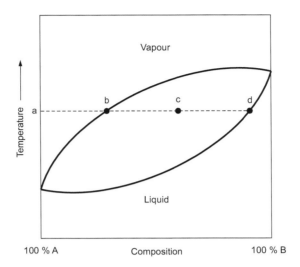

Fig. 3.4: Phase Diagram for a system forming Ideal Solution

In the above system *bcd* is taken as tie line.

$$\frac{\text{Amount in vapour phase}}{\text{Amount in liquid phase}} = \frac{ad - ac}{ac - ab} = \frac{cd}{cb}$$

i.e., Ratio of the amount of two phases is given by the ratio of opposite segments of tie line.

3.5 PHASE DIAGRAM OF TWO COMPONENT SYSTEMS: TYPE A

3.5.1 Only the Pure Components Crystallise from the Solution-Simple Eutectic Diagram:

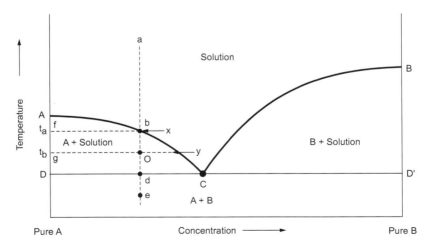

Fig. 3.5: Phase Diagram of a system exhibiting crystallisation of pure components

- **Curves:**

1. Curve AC: Freezing point curve of solid A. Along this curve, solid A is in equilibrium with liquid.

$$\therefore \qquad F' = C - P + 1$$

$$= 2 - 2 + 1 = 1 \text{ (univariant)}$$

2. Curve BC: Freezing point curve of solid B. Along this curve, solid B is in equilibrium with liquid and system is univariant.

- ### *Eutectic Point C*

The two curves AC and BC meet together at point C where there are three phases in equilibrium and system is ($F = 2 - 3 + 1 = 0$) invariant. This point is referred as eutectic point and the corresponding temperature is eutectic temperature.

Eutectic point is a point at which liquid mixture will freeze and represent the lowest melting point of any mixture of two solids.

- ### *Areas*

1. Area $ACDA$: Any point within this area represents solid A in equilibrium with liquid and system is univariant.

2. Area $BCD'B$: Any point within this area represents solid B in equilibrium with liquid and system is univariant.

3. Area above ACB: Here the two components A and B are present as liquid solutions of varying composition. Since the homogeneous phase represents one phase only

$$\therefore \qquad F' = C - P + 1 = 2 - 1 + 1 = 2 \text{ and system here is bivariant.}$$

4. Area below DCD': Represents a solid mixture of A and B. Here there are two phases

$$\therefore \qquad F' = C - P + 1 = 2 - 2 + 1 = 1 \text{ and system is univariant.}$$

Suppose a liquid mixture of composition a is cooled the temperature will fall without any change of composition until point b is reached. At this temperature which corresponds to t_a, solid A starts separating. As cooling continues, A keeps on separating and solution becomes relatively richer in B. Thus, at temperature t_a solid A is in equilibrium with solution of composition x and at t_b it (solid A) is in equilibrium with solution of composition y. The point y is the intersection point of tie-line drawn from point O with the curve AC.

Tie-line is a line which connects different phases in equilibrium with one another. In present case, two phases in equilibrium are represented by points g and y, where g represents solid phase A and y represents liquid phase. The relative amounts of two phases may be determined using lever rule:

$$\frac{\text{Amount of solid } A}{\text{Amount of liquid phase of composition } y} = \frac{yO}{Og}$$

When the eutectic point d is reached, solid B commences to deposit and now the temperature must remain constant until all the liquid is solidified (at d). This results in complete arrest in the rate of cooling, and when the whole liquid has solidified the temperature falls further. On further cooling (d to e), A and B separate out together in a fixed ratio so that their composition remains constant. If the cooling line is drawn to the right of C then solid B will separate out first.

Example of Simple Eutectic Phase Diagram

Phase Diagram of Pb-Ag System

The system has two components: *Pb* and *Ag*.

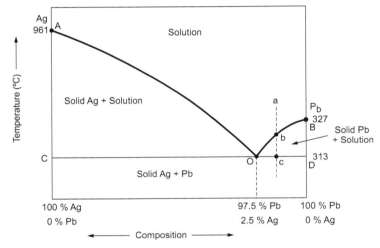

Fig. 3.6: Phase Diagram of Pb-Ag system

The diagram has:

    1. Two curves: AO and BO.

    2. One Eutectic point O.

    3. Three areas

        (*i*) Area above AOB (*ii*) Area AOC

        (*iii*) Area BOD (*iv*) Area below COD.

- **Curves:**

Curve AO: Freezing point curve of Ag. The curve falls off with the addition of Pb to Ag. Along this curve, the two phases solid Ag and solution are in equilibrium.

$$\therefore \qquad\qquad F' = C - P + 1$$
$$= 2 - 2 + 1 \Rightarrow 1$$

and the system Ag/Solution is monovariant.

Curve *BO*: Freezing point curve of Pb. Along this curve Pb and solution are in equilibrium and system is monovariant.

- **Eutectic point *O***: The two curves *AO* and *BO* intersects at point *O*. Here, solid Ag and Pb are in equilibrium with solution. Now, there are three phases:

$$F' = C - P + 1$$
$$= 2 - 3 + 1 \Rightarrow 0$$

and system solid Ag/Pb/Solution is nonvariant. At point *O*, both temperature and composition are fixed (313°C, 97.5% Pb and 2.5% of Ag).

- **Areas**

Area above *AOB*: It consists of single phase system *i.e.*, solution.

∴ Therefore, $F' = C - P + 1$
$$= 2 - 1 + 1 = 2$$

and system is bivariant.

Area within *AOC*: Any point within this area represents solid Ag in equilibrium with solution and system is monovariant.

Area *BOD*: Any point within this area represents solid Pb in equilibrium with solution and system is monovariant.

Area below *COD*: Any point below line *COD* represents two solids Pb and Ag. Now, as there are two phases:

$$F' = C - P + 1$$
$$= 2 - 2 + 1 = 1 \text{ (monovariant)}$$

Pattinson's Process for Desilverisation of Lead

The phase diagram of Pb–Ag system has a special significance in desilverisation of lead. The process of heating argentiferrous lead containing a very small amount of silver (0.1%) and cooling to get pure lead and liquid richer in silver is Pattinson's process.

The argentiferrous lead is first heated to a temperature well above the melting point of pure lead so that system consists only of liquid phase represented by point *a*. It is then allowed to cool and the temperature of the melt falls along line *ab*. At point *b*, solid Pb starts separating. On further cooling, more and more Pb separates and the liquid in equilibrium with solid lead gets richer in silver. Laed continues to separate out and is constantly removed by ladels. The melt continues to be richer in silver until point *O* is reached where percentage of silver has become 2.5. Thus, the original argentiferrous lead which might have contained 0.1% or even lesser amount of silver now contains 2.5% of Ag. So the relative proportion of silver has been raised using Pattinson's process.

3.5.2 The Pure Components Crystallise from the Solution and One of the Solid Exists in more than One Crystalline Form

If one of the component has two crystalline forms α and β.

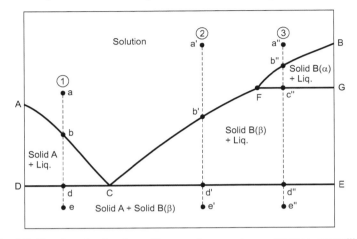

Fig. 3.7: Simple eutectic phase Diagram for a system exhibiting crystalline modification of one component

[Lines 1, 2 and 3 in the phase diagram are referred to as Isopleths]

* **Curves:**

  Curve AC: Along this curve, solid A and liquid are in equilibrium and system is monovariant.

  Curve BC: This curve is further divided into two parts (or curves):

  Curve BF Curve FC

  (*i*) **Curve BF:** Freezing point curve B (α) form. Along this curve B (α) is in equilibrium with solution.

   As there are two phases, system is monovariant.

   $$F' = C - P + 1$$
   $$= 2 - 2 + 1 = 1$$

   The separation of $B(\alpha)$ is continued until point F is reached.

  (*ii*) **Curve FC:** Along this curve, B (β) and solution are in equilibrium and system is monovariant.

  Point F: Transition point where α form changes into β form. At this point $B(\alpha)$, $B(\beta)$ and liquid are in equilibrium and hence system at F is invariant.

   $$F' = C - P + 1$$
   $$= 2 - 3 + 1 = 0$$

- **Areas:**

Area above *ACFB*: Here the single phase system *i.e.*, solution is present and hence system is bivariant.

$$F' = C - P + 1$$
$$= 2 - 1 + 1 = 2$$

Area below *DCE*: Here the system represents solid A in equilibrium with solid B (β) and system is monovariant.

- **Eutectic point *C***: Two curves AC and FC meet at point C where solid A, solid B (β) and solution are in equilibrium and system is invariant.

$$F' = 2 - 3 + 1 = 0$$

- **Cooling Patterns:**

Isopleth: An isopleth is a line in a phase diagram indicating the same composition or mole fraction.

Isopleth-1:

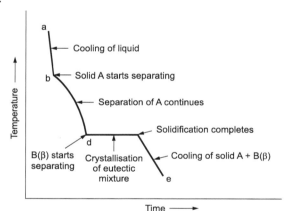

Isopleth-2:

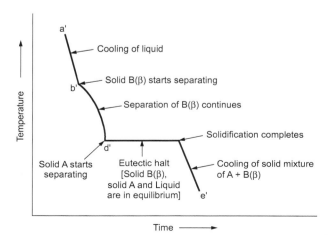

Isopleth-3:

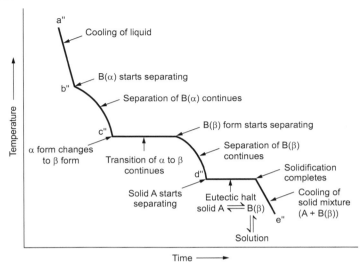

3.5.3 A Solid Compound Stable upto its Melting Point is Formed by Two Constituents

Say we have two components A and B and they form a stable compound A_mB_n.

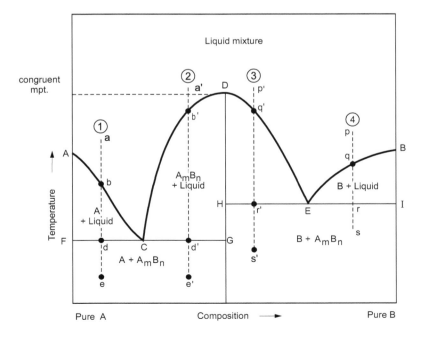

Fig. 3.8: Phase Diagram of a system exhibiting a congruent melting point

- **Curves:**

Curve AC: Along this curve solid A and liquid are in equilibrium and system is monovariant.

Curve BE: Along this curve solid B and liquid are in equilibrium and system is monovariant.

Curve CDE: This is the freezing point curve of solid compound A_mB_n along which A_mB_n and liquid are in equilibrium. At maximum point D on this curve the solid compound and liquid in equilibrium have the same composition and temperature corresponding to D, is the congruent melting point of compound A_mB_n. *Congruent melting point is therefore defined as the temperature where solid and liquid phase of same composition can co-exist. It will be noted that at this point two-component system has virtually become one-component system.*

At D solid A_mB_n and liquid have same composition *i.e.*, $C = 1$.

$\therefore$ $$F' = C - P + 1 = 1 - 2 + 1 = 0 \quad \text{(invariant)}$$

- **Eutectic Point C:** At this point solid A, A_mB_n and liquid are in equilibrium and hence system is invariant.
- **Eutectic point E:** At this point solid B, A_mB_n and liquid are in equilibrium and system is invariant.
- **Areas:**

Area above $ACDEB$: Here the system represents single phase system *i.e.*, solution and hence the system is bivariant.

Area below FCG: The system below FCG represents solid A in equilibrium with solid A_mB_n ($F = 1$).

Area below HEI: The system below HEI represents solid B in equilibrium with A_mB_n ($F = 1$).

- **Cooling Patterns:**

(1)

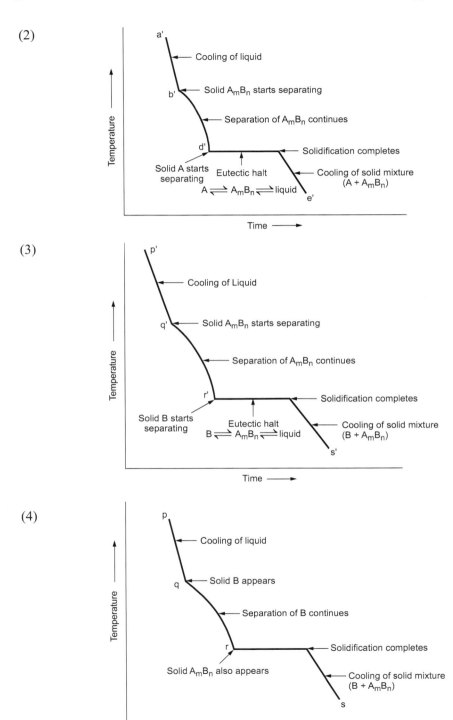

(2)
Cooling of liquid
Solid A_mB_n starts separating
Separation of A_mB_n continues
Solidification completes
Solid A starts separating
Eutectic halt
Cooling of solid mixture $(A + A_mB_n)$
$A \rightleftharpoons A_mB_n \rightleftharpoons$ liquid

(3)
Cooling of Liquid
Solid A_mB_n starts separating
Separation of A_mB_n continues
Solidification completes
Solid B starts separating
Eutectic halt
Cooling of solid mixture $(B + A_mB_n)$
$B \rightleftharpoons A_mB_n \rightleftharpoons$ liquid

(4)
Cooling of liquid
Solid B appears
Separation of B continues
Solidification completes
Solid A_mB_n also appears
Cooling of solid mixture $(B + A_mB_n)$

Note: In some cases, two components form more than one compound. In such cases phase diagram will look like this:

$$eg; A - A_mB, A_mB - AB_n, AB_n - B$$

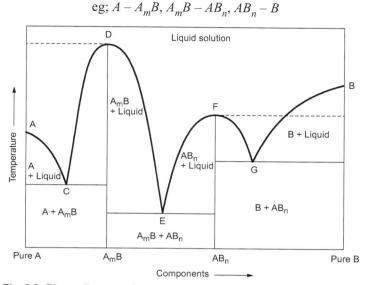

Fig. 3.9: Phase diagram of a system forming two congruent compounds

- **Curves:**

  Curve AC: It represents the equilibrium condition between solid A and liquid solution.

  Curve BG: It represents the equilibrium condition between solid B and liquid solution.

  Curve CDE: It represents equilibrium between solid compound A_mB and liquid. The maximum point D on this curve is the congruent melting point of compound A_mB.

  Curve EFG: It represents equilibrium between solid compound AB_n and liquid. The maximum point F on this curve is the congruent melting point of compound AB_n. At this point the composition of both the phases is identical and therefore $F = 0$.

- **Eutectic points**

  Point C: At this point solid A, solid A_mB are in equilibrium with liquid. Hence, $F' = C - P + 1 = 2 - 3 + 1 = 0$ (invariant).

  Point G: At this point solid B and solid AB_n are in equilibrium with liquid and system is invariant.

  Point E: At this point solid A_mB and solid AB_n are in equilibrium with liquid, $F = 0$.

3.5.4 A Solid Compound Decomposes before it reaches its Melting Point is Formed by Two Constituents

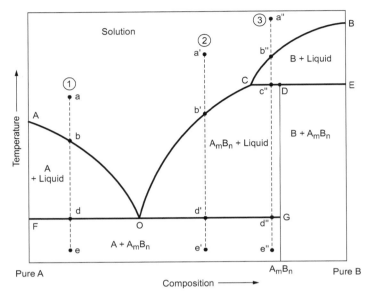

Fig. 3. 10: Phase diagram of a system exhibiting the formation of an incongruent compound

In some cases the compound formed by combination of two components is not stable up to its melting point and it decomposes before reaching the true melting point into another solid, whose composition is different from that of the compound, and a solution. The compound is said to have an incongruent melting point or peritectic point. The situation may be represented as:

$$A_mB_n \rightleftharpoons \text{Solution} + S_{new} \qquad (S_{new} \text{ is a new solid which may be itself a compound or a pure component } A \text{ or } B.)$$

The compound is said to undergo a transition or peritectic reaction.

• **Curves:**

Curve AO: Freezing point curve of solid A. Along this curve, solid A is in equilibrium with liquid and system is monovariant ($F' = 1$).

Curve BC: Freezing point curve of solid B. It represents the equilibrium between solid B and liquid and system here is monovariant ($F' = 1$).

Curve CO: Freezing point curve of compound A_mB_n. It represents the equilibrium between compound A_mB_n and liquid, and system is monovaraint ($F' = 1$).

Point O: Point O is the eutectic point, where solid A, solid A_mB_n and liquid exists in equilibrium and system here is invariant ($F' = 0$).

Point C: At point C, solid B reacts with the melt to form the compound A_mB_n.

$$\text{Solid } B + \text{Solution} \rightleftharpoons \text{Solid } A_mB_n$$

Point C is therefore, also called as transition point. Since, the system will now have three phases *i.e.*, solid B, solid A_mB_n and liquid, therefore, system at point C is invariant. The temperature corresponding to point C is the incongruent melting point or peritectic temperature or transition temperature of compound A_mB_n.

- **Areas:**

  Area *AFOA*: Any point within this area represents equilibrium between solid A and liquid, and system is monovariant.

  Area *BCEB*: Any point within this area represents equilibrium between solid B and liquid, and system is monovariant.

  Area *COGDC*: Any point within this area represents equilibrium between solid A_mB_n and liquid, and system is monovariant.

  Area below *DE*: Any point below line DE represents equilibrium between solid B and solid A_mB_n, and system remains monovariant.

  Area below line *FOG*: Any point below this line represents equilibrium between solid A and solid A_mB_n, and system is monovariant.

  Area above *AOCB*: Area above $AOCB$ represents single phase system *i.e.*, solution and system is bivariant.

- **Cooling Patterns**:

For Isopleth-1:

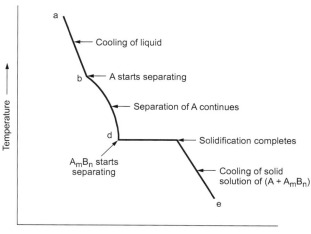

For Isopleth-2:

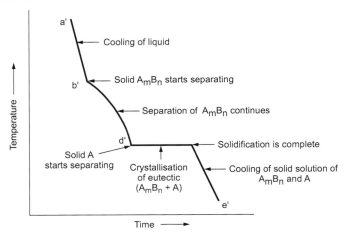

For Isopleth-3:

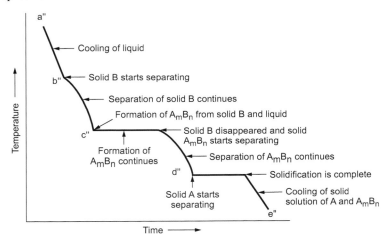

3.5.5 The Two Components are Completely Miscible in the Solid Phase and Form a Series of Solid Solutions

(*a*) The freezing-point curve has neither maximum nor minimum.

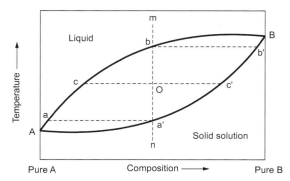

Fig. 3. 11: Phase Diagram of a system exhibiting continuous series of solid solutions with no maximum or minimum

Description of phase diagram:

- **Curves:**
  Curve *AbB*: Freezing point curve of solid solution. (Liquidus curve)
  Curve *Aa'B*: Fusion point curve of solid solution. (Solidus curve)
- **Areas:**
  Area *AbBa'A*: Any point inside this area represents equilibrium between solid solution and liquid.
  Area above *AbB*: Liquid.
  Area below *Aa'B*: Solid solution.
- **Cooling Pattern (Isopleth *mboa'n*):**

If the liquid *m* is cooled, solidification starts at *b* and the composition of solid phase separating corresponds to *b'*. On further decreasing the temperature, solidification is continued and the composition of liquid and solid solution in equilibrium is given by tie-line *coc'* using lever rule:

$$\frac{\text{Amount of solid}}{\text{Amount of liquid}} = \frac{Oc}{Oc'}$$

With the decrease of temperature amount of solid increases relative to that of liquid, and at point *a'* solidification is complete. Separation of solid thus starts at *b* and is complete at *a'* and for this reason freezing curve is also named as **liquidus curve** and the fusion curve is named as **solidus curve**.

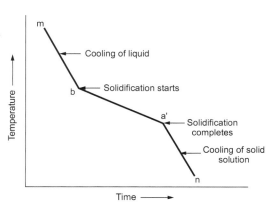

Example of systems showing such type of phase diagrams:

Co – Ni, Au – Ag, Au – Pt and AgCl – NaCl.

(*b*) Freezing-Point Curve has a minimum.

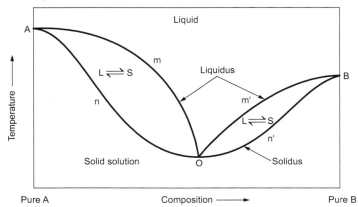

Fig. 3.12: Phase Diagram of a system exhibiting continuous series of solid solutions with a minimum freezing point

- Description of the phase diagram:

A: Freezing point of *A*.

B: Freezing point of *B*.

O: Minimum freezing point at which solid and liquid in equilibrium have the same composition.

Area *AmOnA*: Any point within this area represents equilibrium between solid solution (composition lies on *AnO*) and liquid solution (composition lies on *AmO*).

Area *Bm'On'B*: Any point within this area represents solid solution (composition lies on *Bn'O*) in equilibrium with liquid solution (composition lies on *Bm'O*).

Area above *AmOm'B*: Liquid.

Area below *AnOn'B*: Solid solution.

Examples: KCl – KBr, Cu – Au, Ag – Sb, Mn – Ni, KCl – AgCl etc.

(*c*) Freezing point curve has a maximum.

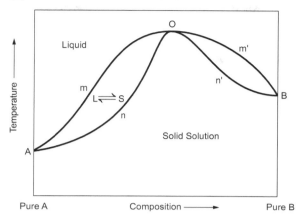

Fig. 3.13: Phase Diagram of a system exhibiting continuous series of solutions with a maximum freezing point

- Description of the phase diagram:

A: Freezing point of *A*.

B: Freezing point of *B*.

O: Maximum freezing point at which solid and liquid in equilibrium have the same composition.

Area *AmOnA*: Any point within this area represents solid solution (composition lies on *AnO*) in equilibrium with liquid (composition lies on *AmO*).

Area *Bm'On'B*: Any point within this area represents equilibrium between solid solution (composition lies on *Bn'O*) and liquid (composition lies on *Bm'O*).

Area above *AmOm'B*: Liquid.

Area below *AnOn'B*: Solid solution.

Examples: d – and l – monobornyl esters of malonic acid, and d – and l – carvoxime.

3.5.6 In the Solid State, the Two Constituents are Partially Miscible and Form Stable Solid Solutions

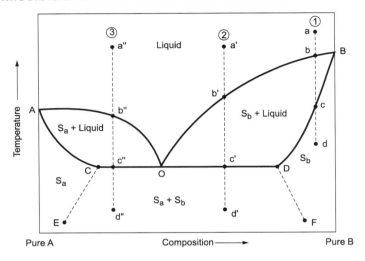

Fig. 3.14: Phase Diagram of a system exhibiting partially miscibility in solid state with eutectic point

- Description of the Phase Diagram:

  Point A: Freezing point of A.

  Point B: Freezing Point of B.

  Curve AO: Freezing point curve of solid solution S_a.

  Curve AC: Fusion point curve of solid solutions S_a.

  Area to the left of ACE: Solid Solution S_a.

  Area AOCA: Any point inside this area represents equilibrium between solid solution S_a with liquid.

  Curve BO: Freezing point curve of solid solution S_b.

  Curve BD: Fusion point curve of solid solution S_b.

  Area to the right of BDF: Solid solution S_b.

  Area BODB: Any point inside this area represents equilibrium between solid solution S_b with liquid.

  Point O: Eutectic point where solid solutions S_a and S_b corresponding to compositions C and D exists in equilibrium with liquid having composition O.

  Area below COD: Two solid solutions S_a and S_b are in equilibrium.

  Area above AOB: Liquid.

  Examples: Au – Ni, Bi – Pb, Fe – Cr, Cd – Sn etc.

- **Cooling Patterns**:

Isopleth-1:

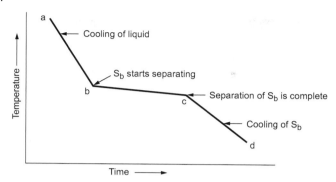

Isopleth-2:

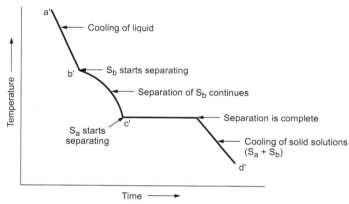

Isopleth-3:

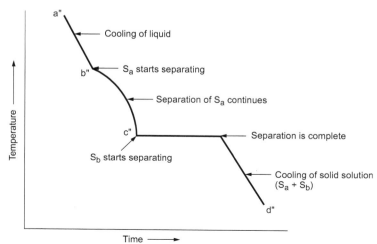

3.5.7 Solid Solutions Formed by Two Constituents and are Stable only upto a Transition Point

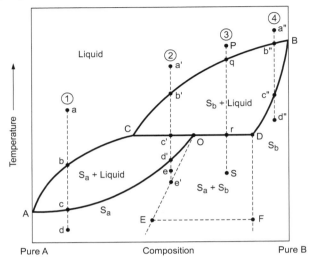

Fig. 3.15: Phase Diagram of a system exhibiting partial miscibility in solid state with a transition point

- Description of the Phase Diagram:

Point A: Freezing point of A.

Point B: Freezing point of B.

Curve AC: Freezing point curve of solid solutions S_a.

Curve AO: Fusion point curve of solid solutions S_a.

Area to the left of AOE: Solid Solution S_a.

Area $AOCA$: Any point within this area represents equilibrium between solid solution S_a with liquid.

Curve BC: Freezing point curve of solid solutions S_b.

Curve BD: Fusion point curve of solid solutions S_b.

Area to the right of BDF: Solid solution S_b.

Area $BCDB$: Any point within this area represents equilibrium between solid solution S_b and liquid.

Point O: Peritectic point or Transition point. (S_b being transformed into S_a at point O.)

Any system at this point (or on line COD) represents equilibrium between solid solution S_a having composition O and solid solution S_b corresponding to composition D with liquid corresponding to composition C.

Area below OD: Two conjugate solid solutions S_a and S_b are in equilibrium with each other.

Area above ACB: Liquid.

- **Cooling Patterns**:

Isopleth-1:

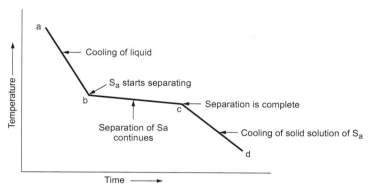

Isopleth-2:

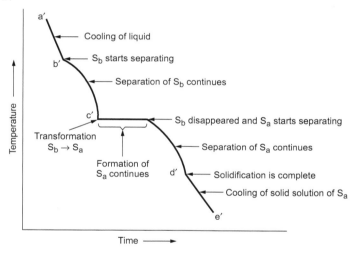

Isopleth-3:

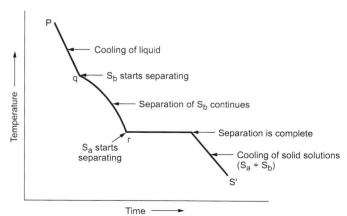

Isopleth-4:

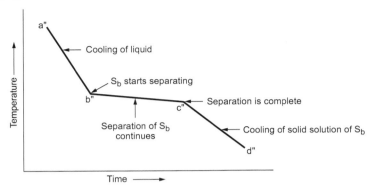

3.6 PHASE DIAGRAM OF TWO COMPONENT SYSTEMS: TYPE B

3.6.1 The two Components are Partially Miscible in the Liquid Phase and Only Pure Components Crystallise from the Solution.

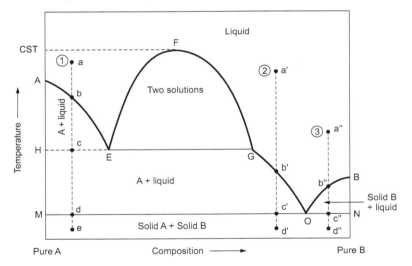

Fig. 3.16: Phase Diagram exhibiting partial miscibility in the liquid phase

- Description of the phase diagram:

    Point A: Freezing point of A.

    Point B: Freezing point of B.

    Curve AE: Freezing point curve of solid A along which solid A and liquid are in equilibrium ($F' = 1$).

    Curve BO: Freezing point curve of solid B along which solid B and liquid are in equilibrium ($F' = 1$).

    Point O: Eutectic point where solid A, solid B and liquid are in equilibrium ($F' = 0$).

Area *AEGOMA*: Any point inside this area represents equilibrium between solid A and liquid ($F' = 1$).

Area *BON*: Any point inside this area represents equilibrium between solid B and liquid ($F' = 1$).

Area below *MON*: Solid A in equilibrium with solid B ($F' = 1$).

Area above *AEFGOB*: Liquid ($F' = 2$).

Point *E*: Solubility limit of B in A is reached at this point and two conjugate layers corresponding to E and G are formed.

Curve *EFGE*: Solubility curve for two partially miscible liquids ($F' = 1$) [Since there are two liquid phases].

Point *F*: Critical solution point.

Examples: Phenol-water, Benzoic acid-water and Benzene-Resorcinol.

- **Cooling Patterns**:

Isopleth-1:

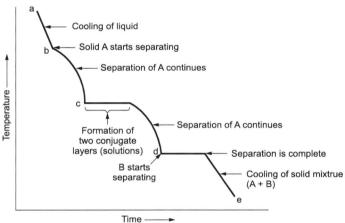

Isopleth-2:

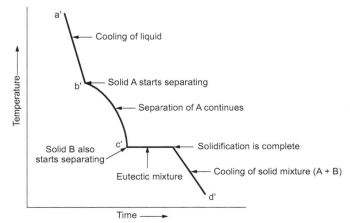

Isopleth-3:

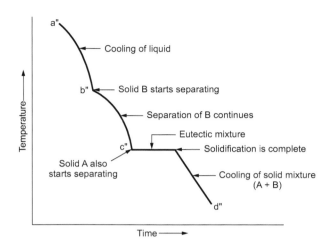

3.7   PHASE DIAGRAM OF TWO COMPONENT SYSTEMS: TYPE C

3.7.1 Formation of Simple Eutectic or Cryohydrates

Potassium iodide-water system (KI-H₂O)

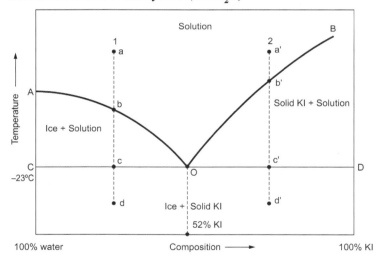

Fig. 3.17: Phase Diagram of KI – H₂O system

- Description of the phase diagram:

Point *A* : Freezing point of water.

Point *AO* : Freezing point curve of water. Along this curve, ice is in equilibrium with solution and hence system is monovariant.

Curve *BO* : Solubility curve of KI. Along this curve, KI is in equilibrium with solution and system is monovariant. This curve depicts the effect of temperature on concentration of saturated KI Solution.

Point O (Eutectic Point): The two curves AO and BO meet at point O, where the solution becomes saturated and KI also separates out as another solid phase. At this point there are three phases *i.e.* solid KI, solid ice and solution and hence system is invariant.

Since, at point O a mixture of ice and KI is deposited which consists of a salt hydrate, (also named as cryohydrate), therefore, eutectic point in this case is also named as cryohydric point.

Area above AOB: Area above AOB represents single phase system *i.e.*, solution and hence system is bivariant.

Area below line COD: Here the system represents ice in equilibrium with solid KI. Now, again there are two phases and hence system is monovariant.

- **Cooling Patterns**:

Isopleth-1:

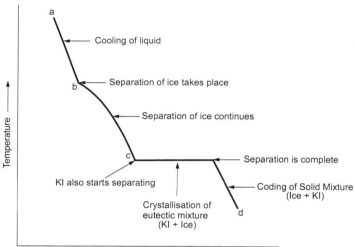

Isopleth-2:

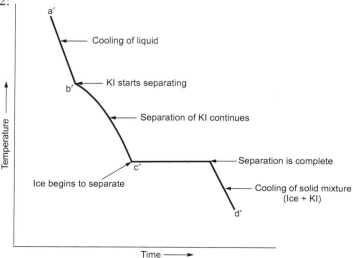

3.7.2 Formation of Compounds (Hydrates) with Congruent Melting Points

Ferric chloride-water system (Fe$_2$Cl$_6$.H$_2$O)

It is a composite system of several simple eutectic systems:

Fe$_2$Cl$_6$ forms four stable crystalline hydrates:

1. Fe$_2$Cl$_6$.12H$_2$O
2. Fe$_2$Cl$_6$.7H$_2$O
3. Fe$_2$Cl$_6$.5H$_2$O
4. Fe$_2$Cl$_6$.4H$_2$O

The other phases are ice, anhydrous Fe$_2$Cl$_6$ and solution.

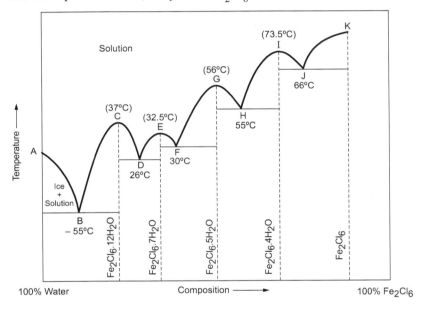

Fig. 3.18: Phase Diagram of Fe$_2$Cl$_6$ – H$_2$O system

- Description of the Phase Diagram:

Point A: Freezing Point of water.

Curve AB: Freezing Point curve of water. Along this curve, there will be fall in temperature with the addition of Fe$_2$Cl$_6$ and this is continued until point B is reached where the dodecahydrate, Fe$_2$Cl$_6$.12H$_2$O appears as a new phase. It represents the equilibrium between ice and solution and system is monovariant.

Point B: Eutectic point where Fe$_2$Cl$_6$.12H$_2$O, ice and solution are in equilibrium and system is invariant ($F' = 0$).

Curve BCD: Solubility curve of dodecahydrate, $Fe_2Cl_6.12H_2O$. It represents the equilibrium between $Fe_2Cl_6.12H_2O$ and solution and system is monovariant.

Point C: At point C, the composition of solution and solid phase ($Fe_2Cl_6.12H_2O$) in equilibrium have the identical composition, therefore C is the congruent melting point of the dodecahydrate.

Curve DEF: Solubility curve of heptahydrate, $Fe_2Cl_6.7H_2O$. At D, the heptahydrate separates out as a new solid phase and hence DEF is solubility curve of heptahydrate. It represents the equilibrium between $Fe_2Cl_6.7H_2O$ and solution and system is monovariant.

Point D: Eutectic point where $Fe_2Cl_6.12H_2O$, $Fe_2Cl_6.7H_2O$ and solution are in equilibrium and system is invariant.

Point E: Congruent melting point of heptahydrate ($F' = 0$).

Curve FGH: Solubility curve of pentahydrate, $Fe_2Cl_6.5H_2O$. It represents the equilibrium between $Fe_2Cl_6.5H_2O$ and solution and system is monovariant.

Point F: Eutectic point where $Fe_2Cl_6.7H_2O$, $Fe_2Cl_6.5H_2O$ and solution are in equilibrium ($F' = 0$).

Point G: Congruent melting point of $Fe_2Cl_6.5H_2O$ where $Fe_2Cl_6.5H_2O$ and solution have the same composition ($F' = 0$).

Curve HIJ: Solubility curve of tetrahydrate, $Fe_2Cl_6.4H_2O$. It represents the equilibrium between $Fe_2Cl_6.4H_2O$ and solution and system is monovariant.

Point H: Eutectic point where $Fe_2Cl_6.5H_2O$, solution and $Fe_2Cl_6.4H_2O$ are in equilibrium ($F' = 0$).

Point I : Congruent melting point of $Fe_2Cl_6.4H_2O$ (tetrahydrate) ($F' = 0$).

Curve JK: At K, the anhydrous Fe_2Cl_6 separates out and therefore it is the solubility curve of anhydrous Fe_2Cl_6 which represents equilibrium between anhydrous Fe_2Cl_6 and solution and system is monovariant.

Point J: Eutectic point where $Fe_2Cl_6.4H_2O$, solution and Fe_2Cl_6 anhydrous are in equilibrium and system is invariant.

3.7.3 Formation of Compounds (Hydrates) with Incongruent Melting Points

Sodium sulphate-water system ($Na_2SO_4 . H_2O$)

Sodium sulphate forms two hydrates:

$$Na_2SO_4.10H_2O \text{ (decahydrate)}$$

and $Na_2SO_4.7H_2O$ (heptahydrate)

Furthermore, the anhydrous salt can exist in two crystalline forms: rhombic and monoclinic. The other phases are solid ice, solution and vapour. Thus, $Na_2SO_4 . H_2O$ is a six phase condensed system.

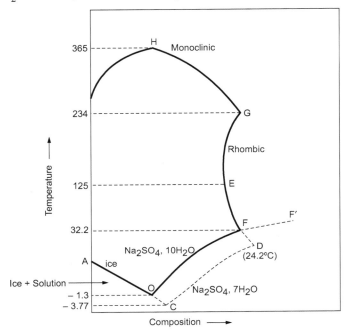

Fig. 3.19: Phase diagram of $Na_2SO_4.H_2O$ system

- Description of the Phase Diagram:

    Curve *AO*: It is the fusion curve of ice. The curve falls off with the addition of anhydrous Na_2SO_4. Along this curve ice and solution are in equilibrium and system is monovariant ($F' = 1$).

    Point *O*: Eutectic point, where a new solid phase *i.e.*, $Na_2SO_4.10H_2O$ appears. At this point $Na_2SO_4.10H_2O$, ice and solution are in equilibrium and system is non-variant ($F' = 0$).

    Curve *OF*: Solubility curve of $Na_2SO_4 . 10H_2O$. If the system at *O* is heated in the presence of excess of salt ice will melt and the salt will dissolve in water thus produced and ultimately one of the solid phase *i.e.*, ice disappears and the travel along curve OF. Along this curve $Na_2SO_4.10H_2O$ and solution are in equilibrium and system is monovariant ($F' = 1$).

    Point *F*: Transition point, where $Na_2SO_4.10H_2O$ changes to rhombic Na_2SO_4 and water:

    $$Na_2SO_4.10H_2O \rightarrow Na_2SO_4 \text{ (rhombic)} + 10H_2O.$$

The temperature corresponding to point F is the incongruent melting point of decahydrate because at this point the composition of liquid phase is different from that of solid phase (hydrate) in equilibrium. Again there are three phases *i.e.*, $Na_2SO_4.10H_2O$, Na_2SO_4 (rhombic) and solution. The system is therefore invariant.

Curve *FEG*: Solubility curve of rhombic Na_2SO_4. If the system at F is heated, all the decahydrate will disappear and the curve FEG is obtained. The system now have only two phases Na_2SO_4 solid (rhombic) and solution, and hence becomes monovariant ($F' = 1$). The curve FEG slopes slightly towards the temperature axis upto 125°C and then slightly away from temperature axis upto 234°C which shows that solubility of rhombic Na_2SO_4 first decreases slightly with increase in temperature (upto 125°C at E) and then increases slightly with further increase in temperature (upto 234°C *i.e.* at G). At G, rhombic Na_2SO_4 changes into monoclinic form. There are three phases in equilibrium *i.e.* Na_2SO_4 rhombic, Na_2SO_4 monoclinic and solution, and system becomes invariant ($F' = 0$).

Point *G*: Polymeric modification of rhombic Na_2SO_4 to monoclinic Na_2SO_4 takes place. At G, three phases namely, rhombic Na_2SO_4, monoclinic Na_2SO_4 and solution exist in equilibrium ($F' = 0$).

Curve *GH*: Solubility curve of monoclinic Na_2SO_4. If heating at G is continued, rhombic Na_2SO_4 changes completely into monoclinic Na_2SO_4. The curve shows that solubility diminishes with increase of temperature upto 365°C which is the critical temperature of solution. Along this curve, Na_2SO_4 monoclinic and solution are in equilibrium and system is monovariant ($F' = 1$). The end point H does not correspond to zero solubility indicating slight solubility of Na_2SO_4 in vapour phase as well.

• *Metastable Equilibrium*

The metastable equilibria of $Na_2SO_4.H_2O$ is shown by dashed lines in phase diagram.

Curve FF': Solubility curve of metastable $Na_2SO_4.10H_2O$. The solubility curve OF of decahydrate is continued beyond the transition temperature upto F'. The significance of this curve is that $Na_2SO_4.10H_2O$ can exist beyond the transition temperature (32.2°C) without changing into rhombic Na_2SO_4.

Curve OC: Freezing point curve of metastable supercooled solution.

The fusion curve AO of ice has been extended along the dotted line OC which shows that it is possible to cool the system even below − 1.3°C without separating $Na_2SO_4.10H_2O$. However, the system will be metastable. In such case, a second hydrate, $Na_2SO_4.7H_2O$ separates out at −3.77°C.

Curve CD: If the temperature is increased at C, the curve CD is obtained which is the solubility curve of metastable $Na_2SO_4.7H_2O$.

Curve FD: Solubility curve of metastable rhombic Na_2SO_4. The curve FEG is extended by dotted line FD. If a solution of rhombic Na_2SO_4 in water is cooled quickly below the transition temperature decahydrate does not appear at F and the curve continues to D.

- **Metastable Eutectic Point C**: At C ($-3.77°C$), another hydrate $Na_2SO_4.7H_2O$ appears. Point C is therefore metastable eutectic point, where ice and $Na_2SO_4.7H_2O$ are in equilibrium with solution ($F' = 0$).

Point D: Point D represents the transition point or incongruent melting point where metastable heptahydrate ($Na_2SO_4.7H_2O$) changes into metastable rhombic Na_2SO_4.

$$Na_2SO_4.7H_2O(s) \rightleftharpoons Na_2SO_4 \text{ (rhombic) + Solution } (D)$$

4

Solutions

4.1 INTRODUCTION

Liquid-Liquid Systems

The solutions of liquids in liquids may be classified as follows:

(*i*) Liquids that are completely miscible

  e.g.: — alcohol and water.

(*ii*) Liquids that are partially miscible

  e.g.: — ether and water, phenol and water.

(*iii*) Liquids that are completely immiscible

  e.g.: — benzene and water.

Solubility of Completely Miscible Liquids

Completely miscible liquids mix in all proportions and hence they could be compared to gases. For such liquids, generally, the volume decreases on mixing but in some cases it increases. Sometimes heat is evolved when they are mixed while in others it is absorbed. The separation of this type of solutions can be effected by fractional distillation.

Solubility of Partially Miscible Liquids

Partially miscible liquids dissolve in each other only to a certain extent *e.g.*, in the case of phenol-water system, phenol will dissolve completely in a large amount of water forming a solution of phenol in water. Now, if the amount of phenol is increased, a stage is reached where saturated solution of phenol in water is formed. Further addition of phenol in water results in the formation of two conjugate layers, one a saturated solution of phenol in water and the other is a saturated solution of water in phenol. If such a system is heated slowly and slowly, more phenol will dissolve and the heterogeneous phases will become homogeneous. For such systems, there is a temperature, at and above which the two components are completely miscible and forms a single solution. This temperature is called critical solution temperature (CST) or consolute temperature which is fixed for every pair of partially miscible liquids.

(*Critical solution temperature is the temperature at which complete miscibility is reached as the temperature is increased.*)

Phase diagram for such a system is constructed by plotting the temperature versus composition of the mixture. Different mixtures of known concentrations are prepared by mixing known amounts of two components. Each mixture is then heated slowly till the heterogeneous phase becomes homogeneous. This temperature is recorded correctly. It is known as Mutual Solubility Temperature (MST) of that mixture. Similarly, MST for each mixture is recorded.

Then a graph is plotted between MST's and composition for various mixtures. The maximum on the curve gives the upper critical solution temperature (UCST).

The relative amounts of two solutions may be calculated using lever rule.

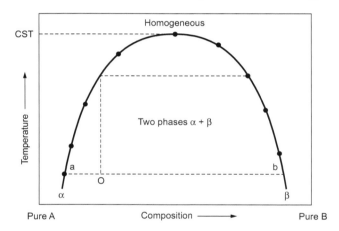

Fig. 4.1: Temperature-composition phase diagram for partially miscible liquid

$$\frac{aO}{bO} = \frac{\text{Amount of substance in solution } \beta}{\text{Amount of substance in solution } \alpha}$$

Outside the curve, the two liquids are completely miscible and form homogeneous solution. *i.e.*, $P = 1$, $C = 2$.

$\therefore$
$$F = C - P + 2$$
$$= 2 - 1 + 2 = 3.$$

Because the pressure is fixed, F reduces to 2 and it is necessary to fix both temperature and concentration to define the system.

Inside the curve, the two layers α and β exist, *i.e.*, $P = 2$, $C = 2$.

$\therefore$
$$F = C - P + 2$$
$$= 2 - 2 + 2 = 2.$$

Again, since pressure is fixed, F reduces to 1 and one variables is to be specified (usually temperature) to define the system completely.

A point on the solubility curve, has one restricting condition of same composition of two solutions. Hence, degrees of freedom $F = (C - r) - P + 2 = (2 - 1) - 2 + 2 = 1$. Now, since P is constant, therefore, $F = 0$ and system is invariant.

The binary systems with upper critical solution temperature (UCST) are

Water + phenol	65.9°C
methanol + n – hexane	33.0°C
methanol + cylohexane	49.1°C
cyclohexane + aniline	29.5°C
aniline + hexane	59.6°C
methanol + n – pentane	14.75°C

Conjugate solutions: *The two solutions having dissimilar composition in equilibrium with each other at a given temperature are known as conjugate solutions.*

4.2 SYSTEMS WITH LOWER CRITICAL SOLUTION TEMPERATURE (LCST)

There are systems in which the solubility decreases with increase of temperature. Such systems have temperature composition curve just reverse of the UCST curve. The system has a lower consolute temperature and therefore, liquids are completely miscible with each other. This type of behaviour is shown by systems consisting of either a hydroxy compound and an amine or a hydroxy-compound and a ketonic group which can from form a hydrogen bond with hydroxyl group.

Phase Diagram of such a system:

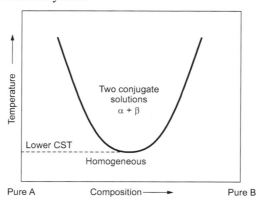

Fig. 4.2: Temperature-composition diagram of a system having lower consolute temperature

Examples of systems showing Lower CST:

triethylamine + water	18.5°C
glycerol + m – toluidine	6.7°C

4.3   SYSTEMS HAVING BOTH UPPER AND LOWER CST

In some cases (or systems), with the increase of temperature, mutual solubility do not continue to decrease. After a certain temperature the solubilities starts to rise again with increase of temperature. Such solutions are expected to show the upper as well as lower critical solutions temperature. For such systems the solubility curve consists of a closed loop. *e.g.*; nicotine-water system with upper CST at 208°C and lower CST at 60.8°C. *i.e.* nicotine and water are completely miscible at a temperature above 208°C and below 60.8°C.

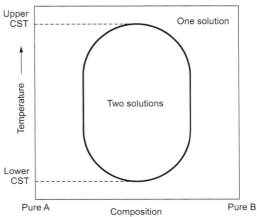

Fig. 4.3: Temperature-composition diagram of a system with both upper and lower CST.

Other systems showing both upper and lower CST:

	UCST/°C	LCST/°C
water + 4 – methylpiperadine	189	86
glycerol + *m* – toluidine	120	6.7
water + β – picoline	153	49

4.4   EFFECT OF PRESSURE ON CST

With the change of external pressure, the CST also varies. On increasing the external pressure, the mutual solubility of two components also increases. Therefore, upper

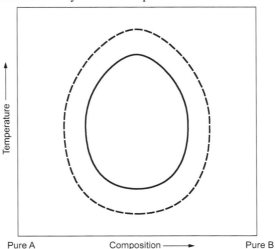

Fig. 4.4: Effect of pressure on CST; Dotted curve at low pressure, Solid curve at high pressure

CST will decrease and lower CST will increase with increase in external pressure. The curve of partial miscibility gets smaller with increase in pressure and finally reduced to a point. At this point the two liquid are completely miscible in each other thereby forming a single solution.

4.5 EFFECT OF IMPURITIES ON CST

The foreign substances *i.e.*, impurities have a marked affect on the critical solution temperature. If the added substance is soluble in one of the liquid, the critical solution temperature increases, due to salting out of water. When the added substance is soluble in both the liquids, it increases the mutual solubilities of the two liquids and critical solution temperature is lowered because of negative salting out effect.

For example, addition of NaCl or naphthalene or KCl increases the CST of phenol-water system. On the other hand, with the addition of succinic acid, CST of phenol-water system is lowered.

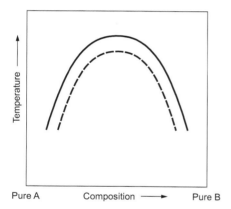
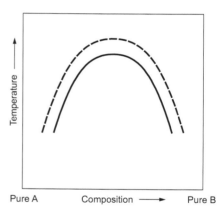

Fig. 4.5 (a) Effect of impurity on CST when the substance is soluble in one of the liquids. Dotted curve — without impurity Thick curve — with impurity

Fig. 4.5 (b) Effect of impurity on CST when the substance is soluble in both liquids. Dotted curve — without impurity Thick curve — with impurity

4.6 VAPOUR PRESSURE-COMPOSITION AND BOILING POINT-COMPOSITION CURVES OF COMPLETELY MISCIBLE BINARY SOLUTIONS

The vapour pressures of two liquids with varying composition have been determined at constant temperature. By plotting the vapour pressure or boiling temperature composition, it has been found that, mixtures of the miscible liquids are of three types:

Type I:

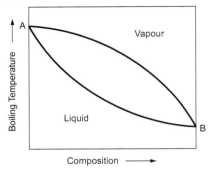
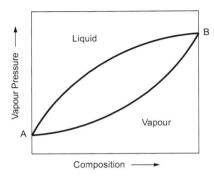

[Fig. 4.6 (a)]

Here the vapour pressure of pure A is lowest and that of B is highest, so the boiling point of A will be highest and that of B the lowest and vapour pressure changes continuously with composition of mixture.

Type II:

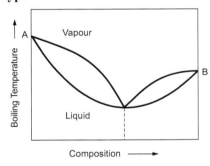
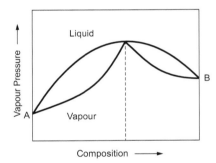

[Fig. 4.6 (b)]

Here the vapour pressure curve shows a maximum for certain composition, the solution of that composition will boil at lowest temperature. This will give rise to a minimum in the boiling point curve.

Type III:

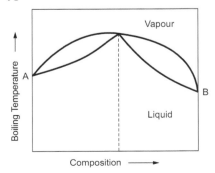
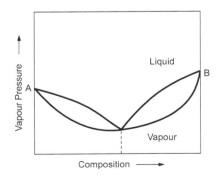

[Fig. 4.6 (c)]

Here the vapour pressure curve shows a minimum for certain composition, the solution of that composition will boil at highest temperature. This will give rise to a maximum in boiling point curve.

4.7 FRACTIONAL DISTILLATION OF BINARY MISCIBLE LIQUIDS

Fractional Distillation is commonly used technique for separating mixture of liquids. It involves repeated vapourisation and condensation of the components of a mixture. Since the boiling point curves of the three types of solutions are different, the behaviour of these solutions on distillation at a constant pressure is also different.

Fractional Distillation of Ideal Solutions

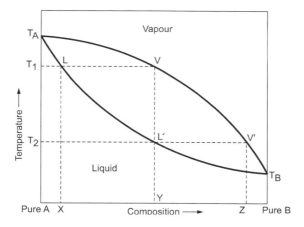

Fig. 4.7: Temperature-Composition diagram for a system forming ideal solution

Suppose a solution of composition X is heated, it boils at temperature T_1. At this temperature the vapour coming off (V) has the composition represented by Y. Since Y is richer in B, the composition of residual liquid will become richer in A which boils at slightly higher temperature. The vapour (V') coming off now has the composition represented by Z. It is still richer in component B and composition of residual liquid is further enriched in liquid A. Hence by repeating the process, boiling point of solution tends to increase (T_1 to T_A) and composition of residual liquid becomes richer in A and hence pure A (lesser volatile constituent) can be obtained as a residual liquid.

Similarly, regarding the vapour phase, if the vapour (V) appeared at T_1 corresponding to composition Y are condensed to form a distillate. This distillate is then heated and the vapour coming off (V') has the composition Z. Distillate is now richer in B than before. Hence, by repeating the process of condensing the vapours to give a new distillate and heating the new distillate, pure B can be obtained as a distillate (more volatile constituent).

This process of separating mixtures by repeated distillation and condensation is extremely tedious. However, this difficulty can be overcome by using fractionating column.

A fractionating column is divided into several parts by means of trays one above the other with a hole in the centre which is covered by bubble cap. Each tray has an overflow pipe that joins it with the tray below by allowing the condensed liquid to flow down.

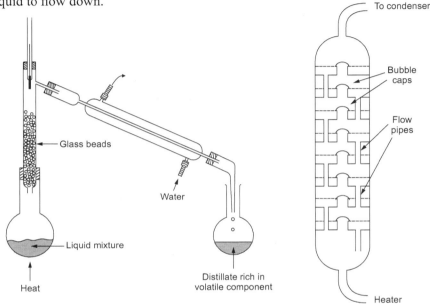

Fig. 4. 8 (a) Fractional distillation with fractionating column

Fig. 4. 8 (b) Fractionating tower for distillation on commercial scale

The fractionating column is fitted in the neck of distillation flask so that the vapours of liquid being heated pass up through it. The temperature falls as the vapours passes from bottom to the top. The hot vapours entering the column condensed first in the lower part of it. With continued heating, more vapours ascend the column and boil the liquid already condensed, giving a vapour that condenses higher up in the column. This liquid is then heated by more vapours ascending the column. Hence, the liquid that is condensed in the lowest part is distilled on to the upper part. In this manner a sort of distillation and condensation takes place along the height of the column resulting in an increase in the proportion of volatile component in the outgoing vapours.

Fractional Distillation of Non-Ideal Solutions

Complete separation of components is possible only in the case of ideal solutions where vapours are richer in more volatile component. In case of non-ideal solutions, the vapour in equilibrium are not always richer in more volatile component. The composition of vapour in this case depends on the type of solution.

(*a*) Fractional distillation of non-ideal solutions showing minimum boiling point.

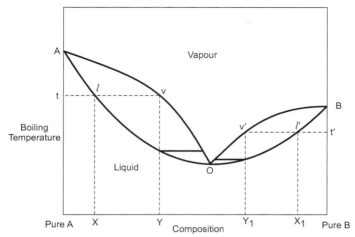

Fig. 4.9: Boiling temperature-composition curve for liquid and vapour phases in non ideal solutions having minimum boiling point

The upper curve indicates the composition of vapour phase and lower one indicates the composition of liquid phase. The two curves meet at minimum point O where liquid and vapour in equilibrium have the same composition. The liquid represented by point O boils at a constant temperature and distil over completely without change of composition. *Such mixtures, which boils at a constant temperature and distils over completely at the same temperature with no change in composition are called* **constant boiling mixtures or azeotropic mixtures.**

Suppose a mixture (liquid) of composition X is heated at temperature t, the vapour coming off has the composition represented by Y. Evidently, it will be richer in constant boiling mixture represented by O. The composition of residual liquid will move towards A. Now, as the distillation proceeds, composition of distillate tends towards O and that of residual liquid towards A. Hence by repeating the process, a constant boiling mixture of composition O will be obtained as the distillate and pure liquid A will be obtained as a residual liquid. It will never be possible to have pure B.

Similarly, if a liquid mixture of composition X_1 is heated, it boils at temperature t', the vapour first evolved has the composition represented by Y_1. Evidently it will be richer in the constant boiling mixture O. The composition of the residual liquid will become richer in B. Now, as the distillation proceeds, the composition of the distillate will approach constant boiling mixture and that of residual liquid will approach pure B. Ultimately, by repeating the process, a constant boiling mixture of composition O will be obtained as the distillate and pure B as the residual liquid. It will never be possible to have pure A.

Therefore, in the fractional distillation of systems having minimum boiling point it is possible to obtain pure A or pure B (if the composition of liquid mixture lies between A and O or between B and O) as a residual liquid and constant boiling mixture of composition O as a distillate.

Examples: ethyl alcohol – water

methyl alcohol – carbon tetrachloride

chloroform – methyl alcohol

(*b*) Fractional distillation of non-ideal solutions showing maximum boiling point.

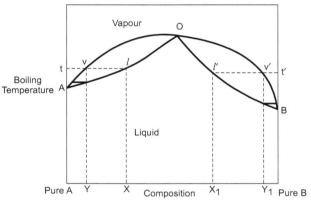

Fig. 4.10: Boiling temperature-composition curve for liquid and vapour phases in non ideal solutions having maximum boiling point

When the liquid l corresponding to composition X is heated, it boils at a temperature t and the vapour first evolved has the composition represented by Y. Evidently, it is richer in A. The composition of residual liquid will move towards O. Now, as the distillation proceeds the distillate will tend towards A and the residue will approach constant boiling mixture O. Hence, by repeating the process, pure A will be obtained as a distillate and constant boiling mixture of composition O will be obtained as a residual liquid. Pure B cannot be obtained at all.

Similarly, if liquid l' corresponding to composition X_1 is heated, it boils at a temperature t', the vapour first evolved has a composition represented by Y_1. Evidently, it is richer in B. The composition of residual liquid will move towards constant boiling mixture O. As the distillation proceeds, the distillate will tend towards B and residue will approach O. Hence, by repeating the process, a distillate of pure B and a residue of constant boiling mixture will be obtained.

Therefore, in the fractional distillation of systems having maximum boiling point, it is possible to obtain pure A or as pure B as a distillate (if composition of liquid mixture lies between A and O or between B and O) and a constant boiling mixture O as a residual liquid.

Examples: acetone – chloroform

water – hydrochloric acid

water – nitric acid.

Hence, in non-ideal solutions fractional distillation results in a separation of a mixture into a residue as a constant boiling mixture and a distillate of either A or B and *vice-versa*. It is never possible to separate a mixture completely into pure A and pure B.

4.8   DISTILLATION OF IMMISCIBLE LIQUIDS

Steam Distillation

The process of steam distillation is used in the purification of substances having high boiling points and are immiscible in water. In this process, steam is passed into the distillation flask containing liquid to be purified. The total vapour pressure over the liquid is now increased, and hence boiling takes place at a lower temperature than the true boiling points of pure liquids.

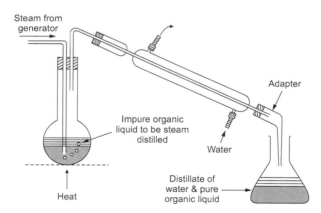

Fig. 4.11: Steam distillation

In this technique, when a mixture of two immiscible liquids is heated, each component independently exerts its own vapour pressure. The total vapour pressure of the system increases. Boiling will start when the sum of the partial pressures of two immiscible liquids A and B exceeds the atmosphere pressure (1 atm.)

$$P_T = P_A + P_B$$

where, P_T = Total vapour pressure of the system

P_A and P_B are the partial pressures of pure liquids A and B respectively.

At the boiling point of mixtures

$$P_A + P_B = 1 \text{ atm.}$$

This temperature would be lower than normal boiling point of either of the liquid. In order to determine the relative amounts of two liquids in distillate, let us assume that n_A and n_B are the number of moles of components A and B. Then,

$$\frac{n_A}{n_B} = \frac{p_A}{p_B}$$

If W_A and W_B are the masses of two components and M_A and M_B are their molar masses, then above equation may be written as:

$$\frac{W_A}{W_B} = \frac{n_A M_A}{n_B M_B} = \frac{p_A M_A}{p_B M_B}$$

Hence, mass of each liquid in distillate is directly proportional to its vapour pressure and molar mass.

Example: An immiscible mixture of water and liquid B boils at 98°C under a pressure of 700 torr. The distillate so obtained contains 70 g of aniline and 1000 g of water. At the given boiling point, the vapour pressure of liquid is 7.0 torr calculate the molar mass of liquid B.

Solutions: Let B stands for liquid B and W for water:

$$p_B = 7 \text{ torr}$$

$$p_W = (700 - 7) \text{ torr} = 693 \text{ torr}$$

$$W_A = 70 \text{ g}$$

$$W_W = 100 \text{ g}$$

Since;

$$\frac{W_W}{W_B} = \frac{p_W M_W}{p_B M_B}$$

$$M_B = \frac{p_W M_W W_B}{W_W p_B}$$

$$= \frac{(693 \text{ torr})(70 \text{ g})(18 \text{ g mol}^{-1})}{(100 \text{ g})(7 \text{ torr})}$$

$$= \frac{873180}{7000}$$

$$= 124.74 \text{ g mol}^{-1}$$

$$M_B = 124.74 \text{ g mol}^{-1}$$

4.9 DUHEM-MARGULES EQUATION

This equation relates the partial vapour pressure of a binary liquid mixture with corresponding mole fractions.

Consider, a binary solution of components A and B respectively, then according to the *GIBBS-DUHEM EQUATION*:

$$n_A\, d\mu_A + n_B\, d\mu_B = 0 \qquad \qquad \text{...(1)}$$

where, μ_A and μ_B are the chemical potentials of A and B, respectively.

$$n_A\, du_A = -n_B\, du_B \qquad \qquad \text{...(2)}$$

Dividing equation (2) by $n_A + n_B$ (total moles of solutions)

$$\frac{n_A}{n_A + n_B}\, du_A = -\frac{n_B}{n_A + n_B}\, du_B$$

$$x_A\, du_A = -x_B\, du_B \qquad \qquad \text{...(3)}$$

where, x_A and x_B are the mole fractions of components A and B respectively.

Dividing equation (3) by dx_A

$$x_A \frac{du_A}{dx_A} = -x_B \frac{du_B}{dx_A} \qquad \qquad \text{...(4)}$$

Since, $\qquad\qquad x_A + x_B = 1$

and, $\qquad\qquad dx_A + dx_B = 0$

$$dx_A = -dx_B$$

Substituting the value of dx_A in equation (4)

$$x_A \frac{du_A}{dx_A} = -x_B \frac{du_B}{(-dx_B)}$$

i.e. $\qquad\qquad x_A \dfrac{du_A}{dx_A} = x_B \dfrac{du_B}{dx_B} \qquad \qquad \text{...(5)}$

The chemical potential of any constituent of a liquid mixture is given by

$$\mu_l = \mu_l^\circ + RT \ln p_l \qquad \qquad \text{...(6)}$$

where $\qquad\qquad p_l =$ partial pressure of the given constituent

$$u_l^\circ = \text{constant}$$

For component A, equation (6) may be written as:

$$\mu_A = \mu_A^\circ + RT \ln p_A \qquad \qquad \text{...(7)}$$

Differentiating equation (7) w.r.t. x_A, keeping temperature and pressure constant.

$$\frac{du_A}{dx_A} = \frac{RTd \ln p_A}{dx_A} \qquad \qquad \text{...(8)}$$

Similarly, for component B;

$$\frac{du_B}{dx_B} = \frac{RT\, d \ln p_B}{dx_B} \qquad \qquad \text{...(9)}$$

Multiplying equation (8) by x_A and equation (9) by x_B.

$$x_A \frac{du_A}{dx_A} = \frac{x_A RT \, d \ln p_A}{dx_A}$$

$$x_A \frac{du_A}{dx_A} = \frac{RT \, d \ln p_A}{d \ln x_A} \qquad \qquad ...(10)$$

Similarly, for the component B,

$$x_B \frac{du_B}{dx_B} = \frac{RT \, d \ln p_B}{d \ln x_B} \qquad \qquad ...(11)$$

Making use of equation (5):

$$\frac{\cancel{RT} \, d \ln p_A}{d \ln x_A} = \frac{\cancel{RT} \, d \ln p_B}{d \ln x_B}$$

$$\boxed{\frac{d \ln p_A}{d \ln x_A} = \frac{d \ln p_B}{d \ln x_B}} \qquad \qquad ...(12)$$

Duhem-Margules equation.

It relates the partial pressures of two constituents with their mole fractions.

4.9.1 Ideal Solutions and the Duhem-Margules Equation

Suppose component A behaves ideally. Then, Raoult's law is applicable for this component,

$$p_A = x_A p_A^o, \text{ where } p_A^o = \text{Vapour pressure of pure liquid } A.$$

Taking logarithms on both sides and differentiating;

$$\ln p_A = \ln x_A + \ln p_A^o$$

$$d \ln p_A = d \ln x_A$$

or $\qquad \qquad \dfrac{d \ln p_A}{d \ln x_A} = 1 \qquad \qquad ...(13)$

Then, according to equation (12), we must have

$$\frac{d \ln p_B}{d \ln x_B} = 1 \qquad \qquad ...(14)$$

Integrating equation (14) gives;

$$\ln p_B = \ln x_B + I \qquad \qquad ...(15)$$

where, I is a constant of integration.

At $x_B = 1$, $p_B = p_B^o$ so that $I = \ln p_B^o$

Hence, $\qquad \qquad \ln p_B = \ln x_B + \ln p_B^o$

$$\ln p_B = \ln (x_B p_B^\circ)$$

∴   or $$\boxed{p_B = x_B p_B^\circ}$$ ...(16)

equation (16) is evidently Raoult's law for component B.

Thus, if in a binary solution, component A obeys Raoult's law, then component B also obey Raoult's law and hence behaves in an ideal manner.

4.9.2 Non-Ideal Solutions and the Duhem-Margules Equation

Non-ideal solutions deviates from Raoult's law and the vapour pressure of the component may be greater or less.

(a) When component A shows positive deviation

$$p_A > p_A^\circ x_A$$ ...(17)

Taking logarithms and differentiating

$$\ln P_A > \ln p_A^\circ + \ln x_A$$

$$\frac{d \ln p_A}{d \ln x_A} > 1$$

Then, according to equation (12), we must have

$$\frac{d \ln p_B}{d \ln x_B} > 1$$

which shows that, $$p_B > p_B^\circ x_B$$

Thus, if in a binary solution, component A shows positive deviation, then component B would also show positive deviation.

(b) When component A shows negative deviation

$$p_A < p_A^\circ x_A$$ ...(18)

Then, $$\frac{d \ln p_A}{d \ln x_A} < 1$$

and we must have,

$$\frac{d \ln p_B}{d \ln x_B} < 1$$ (according to equation (12))

which shows that $p_B < p_B^\circ x_B$

Thus, if in a binary solution component p_A exhibits negative deviation, then component B will also do so.

4.10 KONOVALOV'S RULE

It can be used to study the relation between the compositions of vapour phase in equilibrium with a particular solution (ideal or non-ideal).

From equation (12),

$$\frac{d \ln p_A}{d \ln x_A} = \frac{d \ln p_B}{d \ln x_B}$$

or,
$$\frac{x_A}{p_A} \frac{dp_A}{dx_A} = \frac{x_B}{p_B} \frac{dp_B}{dx_B} \qquad \qquad ...(19)$$

According to Dalton's law of partial pressure,

$$p_A = y_A P$$
$$p_B = y_B P = (1 - y_A)P$$

where, y_A and y_B are the mole fractions of components A and B in vapour phase respectively.

Substituting the above relations in equation (19),

$$\frac{x_A}{y_A P} \frac{dp_A}{dx_A} = \frac{x_B}{y_B P} \frac{dp_B}{dx_B} \qquad \qquad ...(20)$$

Since,
$$x_A + x_B = 1$$
$$dx_A + dx_B = 0$$

and
$$dx_B = - dx_A$$

$\therefore$ Equation (20) be written as,

$$\frac{x_A}{y_A . \cancel{P}} \frac{dp_A}{dx_A} = \frac{-x_B}{y_B . \cancel{P}} \frac{dp_B}{dx_A}$$

$$\frac{x_A}{y_A} \left(\frac{dp_A}{dx_A} \right) + \frac{x_B}{y_B} \left(\frac{dp_B}{dx_A} \right) = 0 \qquad \qquad ...(21)$$

The total vapour pressure is given as

$$P = P_A + p_B$$

Differenting the above equation $w.r.t.$ x_A,

$$\frac{dP}{dx_A} = \frac{dp_A}{dx_A} + \frac{dp_B}{dx_A} \qquad \qquad ...(22)$$

From equation (21),

$$\frac{dp_A}{dx_A} = \frac{-x_B y_A}{y_B x_A} \left(\frac{dp_B}{dx_A} \right)$$

Substituting the value of $\dfrac{dp_A}{dx_A}$ into equation (22)

$$\frac{dP}{dx_A} = \frac{-x_B y_A}{y_B x_A}\left(\frac{dp_B}{dx_A}\right) + \frac{dp_B}{dx_A}$$

$$\boxed{\frac{dP}{dx_A} = \left[1 - \frac{x_B y_A}{x_A y_B}\right]\frac{dp_B}{dx_A}} \qquad \qquad ...(23)$$

The change in total vapour pressure with the addition of component A, i.e. $\left(\dfrac{dP}{dx_A}\right)$ may be positive or negative depending upon the sign of the expression within brackets.

Case I: If $x_B y_A > x_A y_B$ i.e. $\dfrac{dP}{dx_A}$ is positive

i.e. $\qquad\qquad\qquad \dfrac{y_A}{y_B} > \dfrac{x_A}{x_B}$

The ratio of mole fractions of A and B in vapour phase is greater than that in the liquid phase.

In this case, vapour is richer in component A than in liquid phase in equilibruirn with it. Thus the vapour is richer in component whose addition to liquid mixture increases the total vapour pressure (dP/dx_A).

Case II: $x_B y_A < x_A y_B$ i.e., $\dfrac{dP}{dx_A}$ is negative and $\dfrac{dP}{dx_B}$ is positive.

$$\frac{y_A}{y_B} < \frac{x_A}{x_B} \text{ or } \frac{y_B}{y_A} > \frac{x_B}{x_A}$$

i.e. the ratio of mole fractions of A and B in vapour phase is less than that in the liquid phase or the ratios of B to A is greater in vapour phase.

In this case, vapour is richer in component B which when added to the liquid mixture increases the total vapour pressure $\left(\dfrac{dP}{dx_B}\right)$.

So, it can be concluded that the vapour phase is richer in the component whose addition to the liquid mixture increases the total vapour pressure, or the vapour phase is richer in a component which is more volatile. This statement is known as Konovalov's rule.

□□□

Three Component Systems

5.1 INTRODUCTION

For a three-component system, phase rule is:

$$F = C - P + 2$$
$$= 3 - P + 2$$
$$\boxed{F = 5 - P}$$

If $F = 0$, no. of phases would be maximum, hence for invariant system five phases must be present together.

For the complete description of the phase diagram of a three-component system four variables *i.e.*, temperature, pressure and mole fractions of any two components must be known. The phase diagram will be then four-dimensional, which is not easy to draw. Hence, T and P are fixed for a given system and triangular plot as suggested by Stokes and Roozeboom is used to describe the system.

5.2 GRAPHICAL REPRESENTATION

The phase diagram for a three-component system can be represented by an equilateral triangle at constant temperature and pressure.

The sides of the equilateral triangle are divided into 10 equal portions within the triangle and then lines parallel to three sides are drawn.

The concentrations of the three components are represented using the following scheme:

(1) The vertices A, B, C represents the three pure components. For example, point A represents 100% of component A, *i.e.*, at point A; $x_A = 1$, $x_B = 0$ and $x_C = 0$. The vertices labelled as B and C corresponds to pure components B and C, respectively.

(2) The sides of the triangle represents the compositions of different binary systems. The side of the triangle opposite to the corner labelled A, shows the absence of A. Side AB implies the composition of binary system $(A + B)$, where

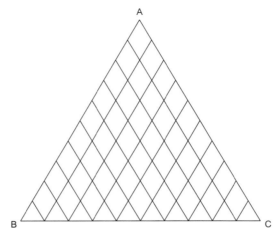

Fig. 5.1: The triangular plot for three component system

the concentration of component C is zero. Side AC represents the composition of binary system $(A + C)$, where concentration of component B is zero. On any line drawn parallel to AB the concentration of C is constant, on lines parallel to AC, concentration of B is constant and on lines parallel to BC, the concentration of A is constant.

(3) Any point within the triangle corresponds to the composition of a mixture of three components.

5.2.1 Method of Calculating the Composition of Each Component

The composition of each component is given by the distance of the point within the equilateral triangle from the sides of the triangle opposite to the respective

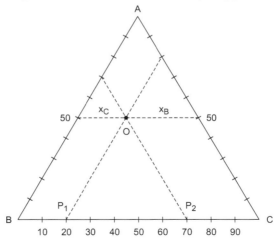

Fig. 5.2: Graphical representation of the composition of a three-component system

apices A, B and C. The distance is then measured along the line parallel to the sides of the triangle.

The mole fraction of A in the mixture (O) is the distance of O from the line BC parallel to AB or AC.

$$x_A = OP_1 = OP_2 = P_1P_2 = 0.5$$
$$x_B = CP_2 = 0.3$$
$$x_C = BP_1 = 0.2$$

5.2.2 Location of the Point within the Triangle when Composition of Each Component is Known

Suppose
$$\left.\begin{array}{l} x_A = a \\ x_B = b \\ x_C = c \end{array}\right\} \text{ Mole fractions of } A, B \text{ and } C.$$
and

The equilateral triangle is drawn and each side is divided into 10 equal parts. A portion $CP_2 = b$ is measured off on line BC which gives the concentration of component B and a portion $BP_1 = c$ is measured off on line BC gives the concentration of component C. The length $P_1P_2 = a$, corresponds to the concentration of the component A. The point of intersection of the parallel lines corresponds to the composition of the ternary mixture.

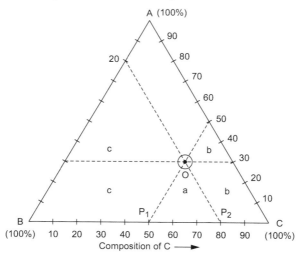

Fig. 5.3: Location of point within the triangular plot

Example: Locate the point inside the equilateral triangle for a ternary mixture with 50% A, 20% B and 30% C.

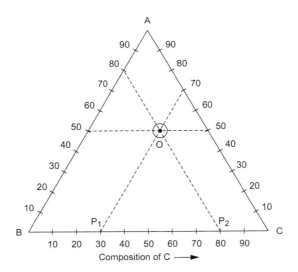

Measure 20% B on line BC as given by CP_2 and 30% C on line BC as given by BP_1. From point P_1, draw a line parallel to side AB and from point P_2 draw a line parallel to side AC. These two parallel lines meet at a point O within the triangle. This point O represents the composition of mixture containing 50% A, 20% B and 30% C.

5.3 SYSTEMS CONSISTING OF THREE LIQUID COMPONENTS EXHIBITING PARTIAL MISCIBILITY

The three liquids A, B and C may constitute the following ternary systems:

(1) Formation of one pair of partially miscible liquids.

(2) Formation of two pairs of partially miscible liquids.

(3) Formation of three pairs of partially miscible liquids.

5.3.1 Formation of one Pair of Partially Miscible Liquids

Let the components be A, B and C. At a given temperature, it may be supposed that liquids A and B are completely miscible, and so also are A and C but B and C are partially miscible.

e.g.: (a) acetic acid, chloroform and water

(b) aniline, phenol and water.

Let the two partially miscible liquids B and C are taken; then there will be two conjugate layers at equilibrium and their compositions may be represented by m and n (A is not present).

Suppose some A is added to the system which is completely miscible with both B and C, it distributes itself between the two layers which now become conjugate ternary solutions. The relative compositions of the two solutions will be given by points b' and c' within the triangle. The points b' and c' may be joined by a tie-line, extremes of which give the composition of two phases in equilibrium. The tie-line $b'c'$ slopes upward to the right indicating that A is relatively more soluble

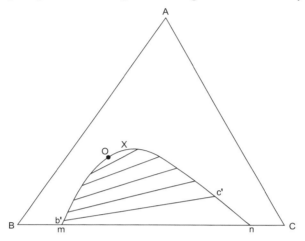

Fig. 5.4: Triangular plot showing the formation of one pair of miscible liquids

in the layer rich in C than in B. Further addition of A not only causes the layers to dissolve more of this component, but the mutual solubilities of B and C are also increased. The composition of two layers approach each other which may be seen by shortening of tie-lines. At point O, the two conjugate solutions have identical composition. Now, the two layers have merged into one. This point is called a '*Plait Point*'.

The various points representing the composition of conjugate layers when joined, a *binodal curve mOXn* is obtained with a maximum at X.

Degrees of freedom in a system exhibiting one pair of partially miscible liquids:

1. Point anywhere outside the Binodal Curve: A point outside the binodal curve *mOXn* represents one liquid layer. Such systems can be completely defined if values of two degrees of freedom, other than temperature and pressure are specified.

$$F = C - P + 2 = 3 - 1 + 2 = 4,$$

$$F = 4 - 2 \qquad \text{(Minus two for temperature and pressure)}$$

$$\boxed{F = 2}$$

2. Point anywhere within the Binodal Curve: A point inside the binodal curve represents two partially miscible liquids ($P = 2$).

$$\therefore \qquad F = C - P + 2 = 3 - 2 + 2 = 3$$

$$F = 3 - 2 \qquad \text{(Minus two for temperature and pressure)}$$

$$\boxed{F = 1}$$

5.3.2   Formation of Two Pairs of Partially Miscible Liquids

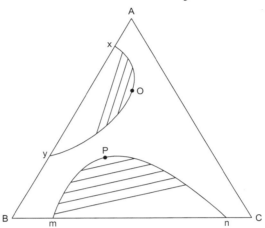

Fig. 5.5: Triangular plot for formation of two pairs of partially miscible liquids

$A - C$ is completely miscible and $A - B$ and $B - C$ are partially miscible. For such systems, there are two binodal curves, one for the system $(A + B) + C$; and other for $(B + C) + A$. The binodal curve mPn represents the behaviour of the mixture of $(B + C) + A$ and P is the plait point. Inside the binodal curve the system exists as two conjugate ternary solutions. The binodal curve xOy represents the behaviour of the mixture of $(A + B) + C$ and O is the plait point. Within the curve the system exists as two conjugate ternary solutions. Outside the binodal curves, the three components A, B and C are completely miscible.

Examples of system exhibiting two pairs of partially miscible liquids:

ethyl alcohol, water and succinic nitrile.

5.3.3   Formation of Three Pairs of Partially Miscible Liquids

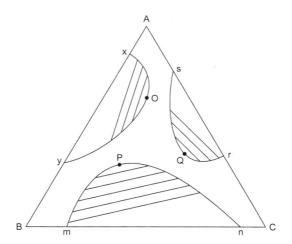

Fig. 5.6: Triangular plot with three partially miscible pairs

For such systems, there are three binodal curves, one for the system $(A + B)$ + C *i.e.*, *xOy*; other for $(B + C) + A$ *i.e.*, *mPn* and one for $(A + C) + B$ *i.e.*, *rQs* with O, P and Q as the respective plait points. Inside the binodal curves, the system exists as two conjugate ternary solutions.

Outside the binodal curves, the three components A, B and C are completely miscible into each other.

e.g.: succinic nitrile, ether and water.

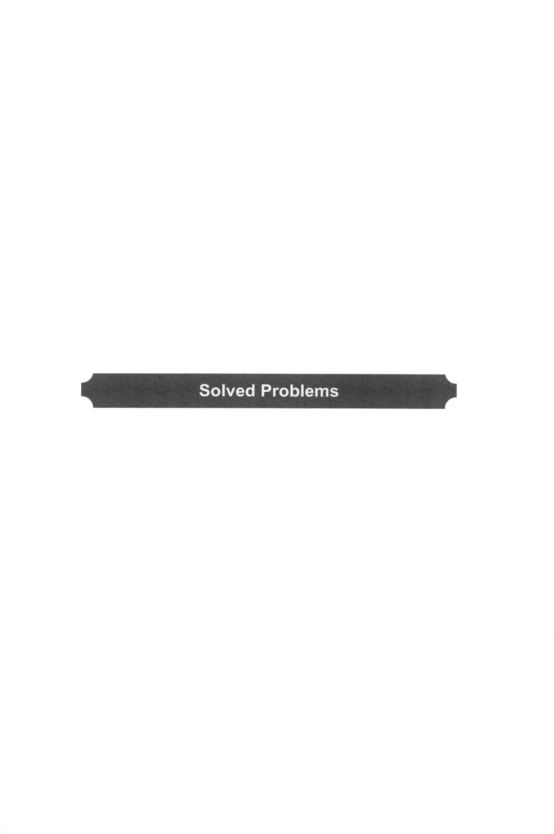

Solved Problems

Question 1: What are the number of components in case of "Two ice cubes" floating on water in a closed container in the presence of water vapours?

Answer: It is a one component system because each phase can be expressed in terms of any one constituent *i.e.* water.

Question 2: Give and justify the number of components in the system?

$$CaCO_3(s) \rightleftharpoons CaO(s) + CO_2(g)$$

Answer: The system consists of three phases – solid $CaCO_3$, solid CaO and gaseous CO_2. Due to the existence of the equation, composition of each of the three phases can be expressed in terms of any of the two constituents – $CaCO_3$, CaO and CO_2. Thus, the number of components = 2. Since, two species out of three are sufficient to express the composition of all the three phases, whichever two are choosen, it is immaterial.

(*i*) If CaO and CO_2 are taken.

$$CaCO_3(s) : CaO(s) + CO_2(g)$$

$$CaO(s) : CaO(s) + OCO_2(g)$$

$$CO_2(g) : OCaO(s) + CO_2(g)$$

(*ii*) If $CaCO_3$ and CaO are taken

$$CaCO_3 : CaCO_3(s) + OCaO.$$

$$CaO(s) : OCaCO_3(s) + CaO(s)$$

$$CO_2(g) : CaCO_3(s) - CaO(s)$$

Thus, it is the number and not the nature that is important for determining the components of the system.

Question 3: Removal of water from a mixture by freeze drying involves cooling below 0°C, reduction of pressure below the triple point and subsequent warming. Explain using phase diagram of water?

Answer: Freeze drying is a dehydration process popularly used in the food and pharmaceutical industries used to preserve a perishable material or make the material more convenient for transport. In this process, the material is cooled below its triple point, the lowest temperature at which the solid and liquid phases of the material can co-exist and subsequently the pressure is lowered and enough heat is supplied to the material for the ice to sublime and converting it directly from the solid phase to the gaseous phase.

The point O in the phase diagram of water where all the three curves OC, OA and OB meet reflect the triple point of water where solid ice, liquid water and vapour are in equilibrium. For water, the point lies at 0.0075°C and 4.6 mm Hg. The curve OB represents the sublimation curve of ice which divides the solid region from the vapour region. For a given temperature the equilibrium between solid and vapour exist only at a specific vapour pressure which can be determined from the curve BO. Freeze drying brings the system around the triple point

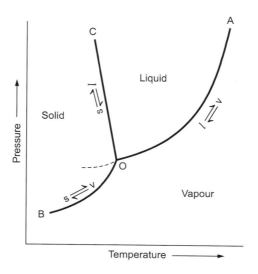

avoiding the direct liquid-gas transition. The system is cooled below its triple point, the pressure is lowered and enough heat is supplied to the material for the ice to sublime. The lowering of pressure lowers the pressure of the vapour phase since at a given temperature, the vapour pressure of solid water has a fixed value, a little bit of ice will sublime to keep the vapour pressure to a constant value. The process is continued till the entire ice sublimes.

Question 4: A substance S can exit in two crystalline forms α and β. The α form is stable under low pressure (whereas the β form is stable under high pressure) condition. The β form is denser than the α-form but both are denser than the liquid. The melting point of α-form is 17°C under a pressure of 8 mm of Hg. The three phases S_α, S_β, S_l are in equilibrium at about 55°C under a pressure of 2000 atm. Sketch a properly labelled diagram of pressure vs temperature.

Given: (1) $S_\alpha \rightarrow$ Stable at lower pressure.

(2) $S_\beta \rightarrow$ Stable under high pressure condition.

(3) Melting point of $S\alpha$ is 17°C at 8mm of Hg.

(4) $P_\beta > P_\alpha$, $P_\beta > P_l$, $P_\alpha > P_l$.

(5) $S_\alpha \rightleftharpoons S_\beta \rightleftharpoons S_l$ at 55°C, 2000 atm.

Answer: Since, high temperature favour the form with greater enthalpy we must have

$$H_{m,\alpha} < H_{m,\beta} < H_{m,l}.$$

Taking point (4) into consideration.

$$P_\beta > P_\alpha > P_l \text{ or}$$

$$V_{m,\beta} < V_{m,\alpha} < V_{m,l}$$

Hence,

$$\left(\frac{dP}{dT}\right)_{\alpha \rightleftharpoons \beta} = \left(\frac{\Delta H_m}{T\Delta V_m}\right)_{\alpha \rightleftharpoons \beta} = \frac{H_{m,\beta} - H_{m,\alpha}}{T(V_{m,\beta} - V_{m,\alpha})} = \frac{+ve}{-ve} = -ve$$

$$\left(\frac{dP}{dT}\right)_{\alpha \rightleftharpoons v} = \left(\frac{\Delta H_m}{T\Delta V_m}\right)_{\alpha \rightleftharpoons v} = \frac{H_{m,v} - H_{m,\alpha}}{T(V_{m,v} - V_{m,\alpha})} = \frac{+ve}{+ve} = +ve$$

$$\left(\frac{dP}{dT}\right)_{\beta \rightleftharpoons l} = \left(\frac{\Delta H_m}{T\Delta V_m}\right)_{\beta \rightleftharpoons l} = \frac{H_{m,l} - H_{m,\beta}}{T(V_{m,l} - V_{m,\beta})} = \frac{+ve}{+ve} = +ve$$

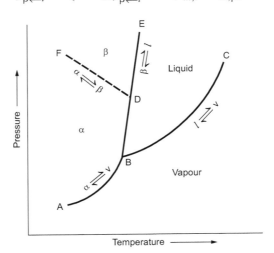

1. Point B: 17°C and 8 mm of Hg where $S_\alpha \rightleftharpoons S_l \rightleftharpoons S_v$.

2. Point D: 55°C and 2000 atm. where $S_\alpha \rightleftharpoons S_\beta \rightleftharpoons S_l$.

Question 5: Explain skating on ice with the help of a phase diagram of water?

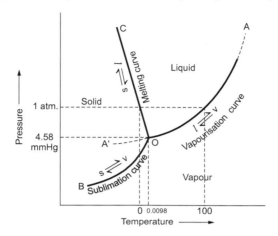

Answer: At the melting point, liquid and solid water co-exist, they are in dynamic equilibrium, since the rate at which ice is melting is just balanced by the rate at which the water is freezing. At 0°C and an external pressure greater than 1 atm, water is a liquid. On phase diagram, ice is subjected to increased pressure at constant temperature. As the pressure is increased, the solid liquid line is crossed, indicating that the ice melts. This is exactly what happens in ice skating. The narrow blades of skates exerts a large pressure, since the skater's weight is supported by the small area on the blade. The ice under the blade melts because of the pressure which provides lubrication. After the blade passes, the liquid refreezes. Without the lubrication effect due to the thawing ice, ice skating wouldn't have been a smooth activity.

Also, mathematically the phenomenon can be expressed using the Clapeyron equation.

$$\left(\frac{dP}{dT}\right)_{s \rightleftharpoons l} = \frac{\Delta H_{m, fus}}{T(V_{m, l} - V_{m, s})}$$

Since, in case of water, $V_{m, s} > V_{m, l}$, the slope $\dfrac{dP}{dt}$ is negative

$$\left(\frac{dP}{dT}\right)_{s \rightleftharpoons l} = \frac{+ve}{-ve} = -ve.$$

Thus, in case of substances such as water for which volume decreases on melting, a decreases in the melting point occurs when the external pressure increases. Hence, during skating, melting point of ice decreases with increase in pressure and therefore skating on ice becomes possible.

Question 6: As heat is removed from a liquid which tends to supercool, its temperature drops below the freezing point and then rises suddenly. What is the source of heat which cause the temperature to rise.

Answer: The enthalpy of fusion.

Question 7: As supercooled water freezes spontaneously, its temperature rises to 0°C. What is the source of heat for the process.

$$H_2O(l) \ (-10°C) \rightarrow H_2O(s) \ (0°C)?$$

Answer: In this case, $\Delta H = 0$. No energy is transferred to or from the system. The energy liberated in the freezing process warms the system to 0°C.

Question 8: Distinguish between the triple point and freezing point of a pure substance.

Answer: The triple point is the point where solid, liquid and vapour are in equilibrium with one another with no other substance present. The freezing point is the point at which solid and liquid are in equilibrium under 1 atm. pressure.

Question 9: For most substances which is higher the triple point or the freezing point?

Answer: The freezing point is higher for most substances which have a positive slope for the solid-liquid equilibrium line in the phase diagram.

Question 10: A substances X has its triple point at 20°C and 0.5 atm, its melting point is 25°C and boiling point is 290°C. Sketch schematic phase diagram for X.

Answer:

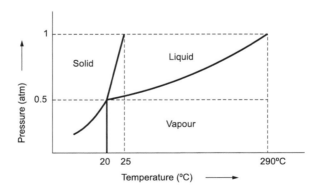

Question 11: Under what conditions the following system is non-variant?

Melting point of a pure substance

Answer: $F = C - P + 2$

For non-variant condition,

$$C - P + 2 = 0$$

since, $C = 1$

$$1 - P + 2 = 0$$

$\boxed{P = 3}$ Required condition.

Question 12: Determine the variance at eutectic point.

Answer: At eutectic point, two components in consideration are in equilibrium with the solution [$C = 2$ and $P = 3$].

Therefore, $F = C - P + 1$

$$= 2 - 3 + 1$$

$$= 0$$

Hence, the variance at eutectic point is zero.

Question 13: Addition of napthalene increases the *CST* of phenol-water system at constant pressure. Explain.

Answer: This is because, the added substance napthalene is soluble only in phenol which is an organic solvent (based on like dissolves like principle) and insoluble in water. This will effect the solubility of water in phenol. In general, the mutual solubilities of the two liquids are decreased and consequently, the *CST* is raised as shown by the dotted curve in the figure.

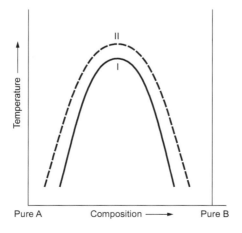

I. CST of two partially miscible liquids (without impurity).

II. CST raised on the addition of a substances soluble in one of the liquids.

Question 14: A congruent melting compound and a eutectic mixture exhibit sharp melting point. How can these two be distinguished?

Answer: We have a rounded maximum in case of congruent melting point. This is because the compound is not much stable at its melting point. It partly dissociates and the products of dissociation in the liquid phase depresses the actual melting point of the compound with the result that the curve has a flattened maximum unlike the eutectic point.

Question 15: It has been found that *CST* of phenol-water is increased by the addition of a small amount of solid KCl. Explain why?

Answer: Addition of a small amount of solid KCl increases the *CST* of phenol-water system. This is because KCl being soluble only in one out of the two liquids (*i.e.*, water) effect the solubility of the other liquid in this liquid, *i.e.*, the solubility of phenol in water is effected. Since KCl is soluble in water and hence a portion of water will be extracted by it from a homogeneous solution of water and phenol. Therefore, the two layers will separate. Thus, the mutual solubilities of the two liquids are decreased and hence the *CST* is raised.

Question 16: A pure compound and a eutectic mixture both show sharp melting point. How would you distinguish between the two?

Answer: A pure compound as well as eutectic mixture both show sharp melting point, but a pure compound has a definite stoichiometric proportion of its constituents unlike the eutectic mixture. The eutectic mixture has a sharp melting point *i.e.*, it melts to give the liquid of the same composition resembling the compound in this sense. A eutectic mixture does not have its components in definite stoichiometric proportions. When the eutectic mixture is observed under a microscope, the components exist as separate crystals revealing its heterogeneity, Also the eutectic mixture has all its components in equilibrium with the solution.

Question 17: Explain that a eutectic mixture has a definite composition and a sharp melting point, yet it is a compound?

Answer: A eutectic mixture has a definite composition and a sharp melting point, yet it is not considered as a compound because of the following reasons:

(*i*) The components present in a eutectic mixture are not in stoichiometric proportions.

(*ii*) When examined under a microscope, the components of a eutectic mixture exist as separate crystals.

Question 18: During thermal analysis the cooling curve of a eutectic mixture has no break point. Explain.

Answer: A binary system consisting of two substances, which are miscible in all proportion in the liquid phase, but which do not react chemically is known as the eutectic (easy to melt) system. At the eutectic temperature, the eutectic mixture will start to solidify. Since, the solid and liquid phases are in same composition, no change in concentration will occur as the solidification takes place. Thus, while cooling a mixture of eutectic composition, solidification of the whole sample takes place at a single temperature. This results in a cooling curve similar to that of a single component system with the system solidifying at its eutectic temperature. When the entire sample has solidified, the temperature will again fall. It is thus impossible to distinguish between a pure substance and a eutectic mixture on the basis of the cooling curve alone.

Question 19: What is an azeotrope? It is a compound?

Answer: An azeotrope boils at a constant temperature and distil over without change in composition at any given pressure like a pure compound. These cannot be regarded as compounds because when the total pressure is changed, both the boiling point and composition of the azeotrope changes whereas for a chemical compound the composition should remain constant.

Question 20: If pure ethanol has a boiling point of 78.3°C and its azeotrope has a boiling point of 78.174°C, what would its graph look like.

Answer: Since the azeotrope boiling point < pure ethanol boiling point, the azeotrope is a positive azeotrope.

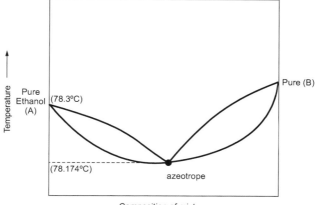

Composition of mixture

Question 21: Why does the plait point lie generally either to the left or the right of the maximum of the binodal curve?

Answer: Plait point is an invariant point, *i.e.*, at this point, the composition of the two solutions become identical and thus two solutions merge into each other and form a single solution. This point generally lies to the either to the left or to the right side of the maximum of the binodal curve, since various tie lines are not horizontal. The tie lines are neither parallel to the line *BC* nor parallel to each other, if the added component *A* is not equally soluble is both the solutions. If the component *A* is relatively more soluble in the layer rich in one component *i.e. C* than it is in the layer rich is another component *i.e. B*, then obviously the state point of the layer rich in *C* former component will lie nearer to the apex *A* than that of the layer rich is *B*.

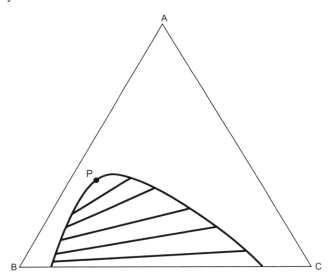

Consequently, the tie-lines joining these two state points will be sloped upward to the right. On the other hand, if the component *A* is more soluble in layer rich in *B*, then tie-lines will be sloped upward to the left.

Hence, the plait point lies, generally, to the either side of the maximum of the binodal curve.

Question 22: What happens when small amount of the component A is gradually added to the binary system of partially miscible component B and C. (All of them are assumed to be in liquid form.)

Answer:

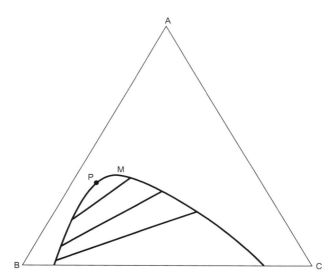

If a component A is added gradually in small amounts to the binary system of partially miscible component B and C, since A is miscible in both the conjugate solution, the addition of A will cause an increase in the mutual solubilities of the two components B and C. Consequently compositions of the two conjugate solutions will now become closer to each other and at point P, two solutions become identical and thus two solutions merge into each other forming a single solution.

Question 23: What is retrograde solubility? Explain with the help of a phase diagram.

Answer: Considering a system containing 3 components A, B and C, let the proportion of component A be constant and the amounts of other two components B and C constituting the binary system be varied.

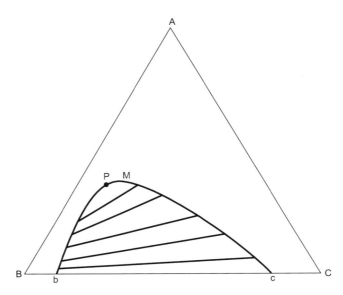

If the original composition of the system happens to lie between P (plait point) and M (maximum of the binodal curve), it is obvious that the single solution that splits into two is of the one of B in C. The two solutions again merge on the other side of the binodal curve to give back the solution of B in C. This behaviour is observed whenever the amount of A in the original system lies between plait point P and the maximum point M. Such solutions are said to show *retrograde solubility.*

Question 24: The plait point in a 3-component system having one partially miscible pair of liquid is an invariant point. Explain.

Answer: At the plait point, composition of two solutions become identical and thus two solutions merge into each other and form a single solution. Therefore, 3 component system becomes 2- component. Hence, F = C - P + 2, = 2 - 2 + 2, = 2

i.e. F = 2 - 2 (minus 2 for T and P) Hence, F = 0.

Thus, the plait point in a 3-component system having one partially miscible pair of liquids is an invariant point.

Question 25: Describe the phenomenon of "salting out".

Area:

AfbA: K_2CO_3 in equilibrium with saturated solution whose composition lies on *fb*.

AceA: K_2CO_3 in equilibrium with saturated solution whose composition lies on *ec*.

byczb: Two conjugate liquids (water and alcohol).

Abca: K_2CO_3 in equilibrium with conjugate liquids of compositions *b* and *c*.

Answer:

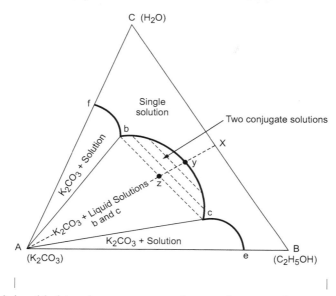

If a salt is added to a homogeneous mixture of water and some other organic liquid component (ethyl alcohol), sometimes the solution separates into two layers. This phenomenon is known as salting out effect. For example, if potassium carbonate is added to a homogeneous solution of water + ethanol, two liquid layers separate; one rich in alcohol and the other being rich in water.

Question 26: The tie lines within a binodal curve of a three component system are parallel neither to the sides of triangle nor to each other. Explain.

Answer:

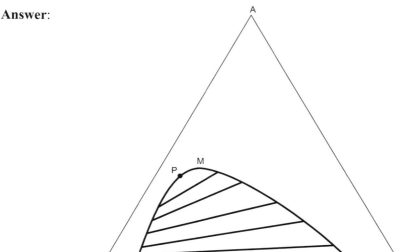

Triangular plot showing the formation of one pair of partially miscible liquids

As usual the composition of the two conjugate solutions will be given by the two ends of tie line. In general the various tie lines are neither parallel to the line BC nor parallel to each other, if the added component A is not equally soluble in both the solutions. If the component A is relatively more soluble in layer rich in C than it is in the layer rich in B, then obviously the state of the point of the layer rich in C will lie near to the apex A than that the layer rich in B. Consequently, the tie-line joining the two state points will be sloped upward to the right. However, if the component A is more soluble in layer rich in B, then various tie-lines will be sloped upward to the left.

Question 27: What is a binodal curve in a three component system having a pair of partially miscible liquid?

Answer:

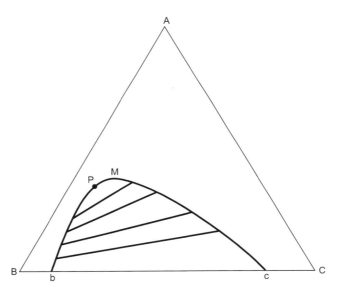

Triangular plot showing the formation of one pair of partially miscible liquids

When a smooth curve passing through various points representing the composition of the conjugate solutions is drawn, one gets a binodal curve bPc, with a maximum at M. In the particular case of a system having one pair of partially miscible liquids, various lie-lines are horizontal and thus at the maximum point P, the two solutions have identical composition and thus merge into each other to form a single solution. Any point outside the binodal curve $bPMC$ represents one liquid only. Any point inside the curve represents a system in which the two partially miscible liquid solutions are formed.

Question 28: Prove that plait point for a three component system with one pair of partially miscible liquids is invariant point.

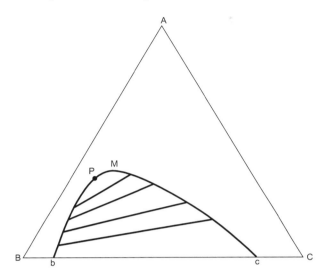

Answer: If the components are *A*, *B* and *C* then it may be supposed that at a given temperature the liquid *A* and *B* are completely miscible and *A* and *C* are also completely miscible but *B* and *C* are only partially miscible.

e.g. $CH_3COOH - CHCl_3 - H_2O$

Suppose the two partially miscible liquids *B* and *C* are taken, then at equilibrium, there will be two conjugate layers whose composition may be explained by points *b* and *c*.

Plait point: In the diagram at point *P*, composition of the two solution become identical and thus two solution merge into each other and form a single solution. At this point the composition of two phases in equilibirm becomes identical.

Hence, C = 2 (instead of 3)

$$F = C - P + 2$$
$$= 2 - 2 + 2$$
$$= \quad 2$$
$$F = 2 - 2 \text{ (minus 2 for T and P)}$$
$$F = 0$$

Hence, the system is invariant.

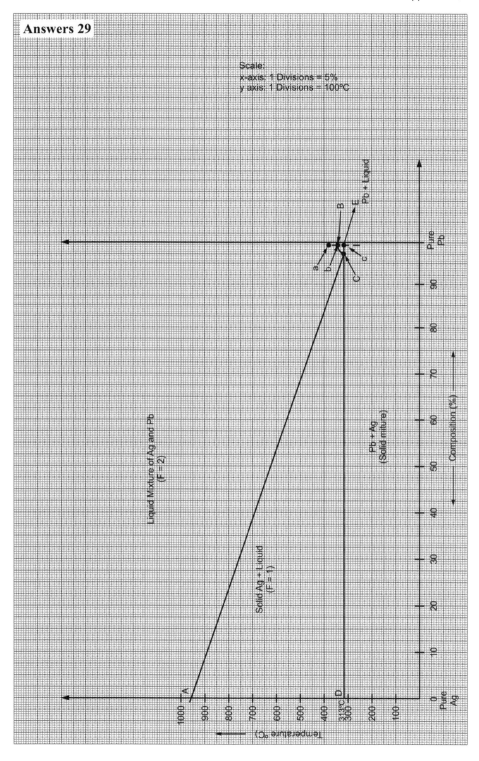

Answers 29

Scale:
x-axis: 1 Divisions = 5%
y axis: 1 Divisions = 100°C

Question 29: Plot a Phase diagram with the following data: M.P. of silver = 961°C, M.P. of lead = 327°C, eutectic point = 313°C, corresponding to 97.5% of lead. Plot the graph and label each of the area and region.

Explain the significance of desilverisation of lead or Pattinson's process using the phase diagram of lead and silver.

Explanation: The process of heating argentiferrous lead containing a very small amount of silver (say 0.1%) and cooling to get pure lead and liquid richer in silver is Pattinson's process.

The argentiferrous lead is first heated to a temperature well above the M.P. of pure lead so that the system consists only of liquid phase represented by point *a*. It is then allowed to cool and the temperature of the melt fall along line *ab*. At point *b*, solid lead starts separating. On further cooling, more and more lead separates out and the liquid in equilibrium with lead becomes richer in silver. The melt continues to be richer in silver until point *C* is reached where percentage of silver has becomes 2.5. Thus, the original argentiferrous lead which might have contained 0.1% (or even less) of silver can now contain 2.5% of silver. So, the relative proportion of silver has been increased using Pattinson's process.

Answers 30

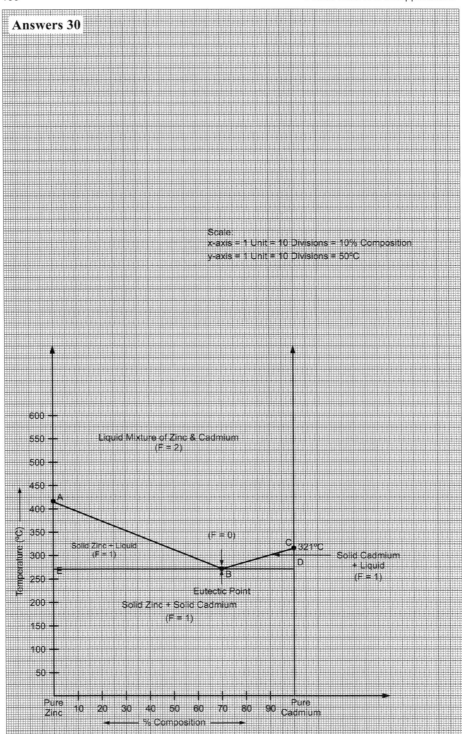

Scale:
x-axis = 1 Unit = 10 Divisions = 10% Composition
y-axis = 1 Unit = 10 Divisions = 50°C

Liquid Mixture of Zinc & Cadmium
(F = 2)

A

Solid Zinc + Liquid
(F = 1)

(F = 0)

C 321°C

Solid Cadmium
+ Liquid
(F = 1)

D

E

B

Eutectic Point

Solid Zinc + Solid Cadmium
(F = 1)

Temperature (°C)

600
550
500
450
400
350
300
250
200
150
100
50

Pure
Zinc   10   20   30   40   50   60   70   80   90   Pure
Cadmium

◄———— % Composition ————►

Question 30: Construct the phase diagram using the following data: M.P. of Zinc = 419°C and that of cadmium = 321°C. Eutectic temperature corresponds to 270°C. Label the graph and explain the salient features.

Explanation:

Point A: Melting Point of zinc.

Point C: Melting Point of cadmium.

Point B: Eutectic point where, liquid $\rightleftharpoons$ solid zinc $\rightleftharpoons$ solid cadmium. System here is invariant.

Line ED: Eutectic line. Solidification of zinc and cadmium takes place and the system here is invariant.

Area ABE: Any point in this area represents solid zinc $\rightleftharpoons$ liquid and the system here is monovariant.

Area CBD: Any point in this area represents solid cadmium $\rightleftharpoons$ liquid and the system here is monovariant.

Area below ED: Any point within this area represents solid zinc $\rightleftharpoons$ solid cadmium and the system here is monovariant.

Area above ABC: Liquid mixture of zinc and cadmium, system here is bivariant.

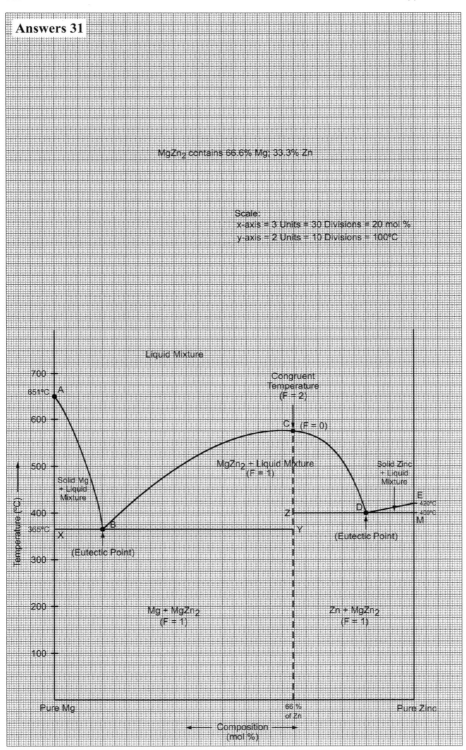

Answers 31

MgZn$_2$ contains 66.6% Mg; 33.3% Zn

Scale:
x-axis = 3 Units = 30 Divisions = 20 mol %
y-axis = 2 Units = 10 Divisions = 100°C

Liquid Mixture

700

651°C A

600   Congruent
                                             Temperature
                                             (F = 2)

                                         C (F = 0)

500   MgZn$_2$ + Liquid Mixture   Solid Zinc
                                    (F = 1)   + Liquid
     Solid Mg   Mixture
     + Liquid   E
     Mixture   420°C
400   D   430°C
365°C   Z   Y   M
     X   B
                                                                 (Eutectic Point)
       (Eutectic Point)

300

200   Mg + MgZn$_2$ Zn + MgZn$_2$
                    (F = 1)   (F = 1)

100

     Pure Mg   66 %   Pure Zinc
                                                  of Zn
                        ←— Composition —→
                              (mol %)

Temperature (°C) →

Question 31: Construct the phase diagram using following data:

- M.P. of Mg is 651°C and M.P. of Zn is 420°C.
- Compound formed $MgZn_2$ melts at 575°C.
- Eutectic is at 365°C, containing Mg and $MgZn_2$ and another at 400°C containing Zn and $MgZn_2$ in equilibrium with liquid.

Explanation:

Point A: Melting point of Mg.

Point E: Melting point of Zn.

Curve AB: Freezing point curve of Mg.

Curve ED: Freezing point curve of Zn.

Point C: Congruent melting point, solid $MgZn_2 \rightleftharpoons$ liquid, $F = 0$.

Point B: Eutectic point, solid Mg $\rightleftharpoons$ solid $MgZn_2 \rightleftharpoons$ liquid, $F = 0$.

Point D: Eutectic point, solid Zn $\rightleftharpoons$ solid $MgZn_2 \rightleftharpoons$ liquid, $F = 0$.

Area ABX: Solid Mg $\rightleftharpoons$ liquid, $F = 1$.

Area $BCDZY$: Solid $MgZn_2 \rightleftharpoons$ liquid, $F = 1$.

Area DEM: Solid Zn $\rightleftharpoons$ liquid, $F = 1$.

Area below XY: Solid Mg $\rightleftharpoons$ solid $MgZn_2$, $F = 1$.

Area above ZM: Solid Zn $\rightleftharpoons$ solid $MgZn_2$, $F = 1$.

Area above $ABCDE$: Liquid mixture of magnesium and zinc.

Line XBY: Eutectic line, a system on this line represents solid Mg and solid $MgZn_2$ in equilibrium with liquid.

Line ZDM: Eutectic line, a system on this line represents solid Zn and solid $MgZn_2$ in equilibrium with liquid.

Answers 32

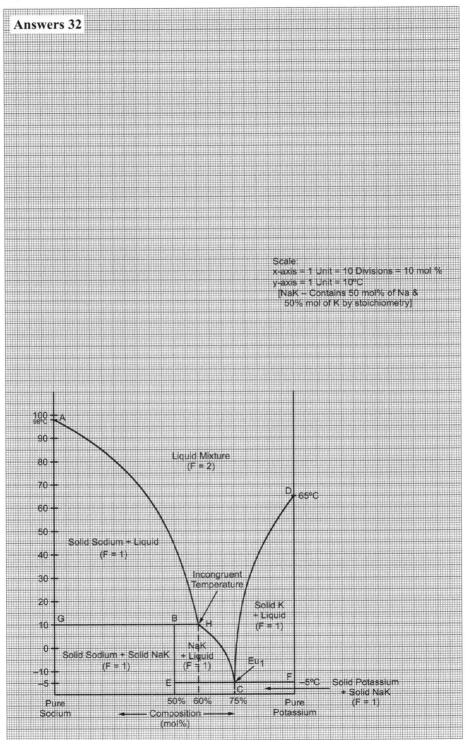

Scale:
x-axis = 1 Unit = 10 Divisions = 10 mol %
y-axis = 1 Unit = 10°C
[NaK – Contains 50 mol% of Na &
50% mol of K by stoichiometry]

Question 32: Sodium and potassium melts at 98°C and 65°C respectively. They form one compound NaK which decomposes at 10°C to give a solid and a melt containing 60 mol% of K. There is an eutectic at –5°C and the eutectic composition is 75 mol% of K. Sketch the phase diagram and label it.

Explanation:

Curve AH: Freezing point curve of Na.

Curve DC: Freezing point curve of K.

Point H: Incongruent melting point, solution $\rightleftharpoons$ solid Na $\rightleftharpoons$ solid NaK, $F = 0$.

Point C: Eutectic point, liquid $\rightleftharpoons$ solid NaK $\rightleftharpoons$ solid K, $F = 0$.

Line EF: Eutectic line, a point in this line represents solid NaK and solid K in equilibrium with liquid.

Area AHG: Solid Na $\rightleftharpoons$ liquid, $F = 1$.

Area DCF: Solid K $\rightleftharpoons$ liquid, $F = 1$.

Area HCEB: Solid NaK $\rightleftharpoons$ liquid, F = 1.

Area below GB: Solid sodium $\rightleftharpoons$ solid NaK, $F = 1$.

Area below EF: Solid potassium $\rightleftharpoons$ solid NaK, $F = 1$.

Area above AHCD: Liquid mixture of Na and K, $F = 2$.

Answers 33

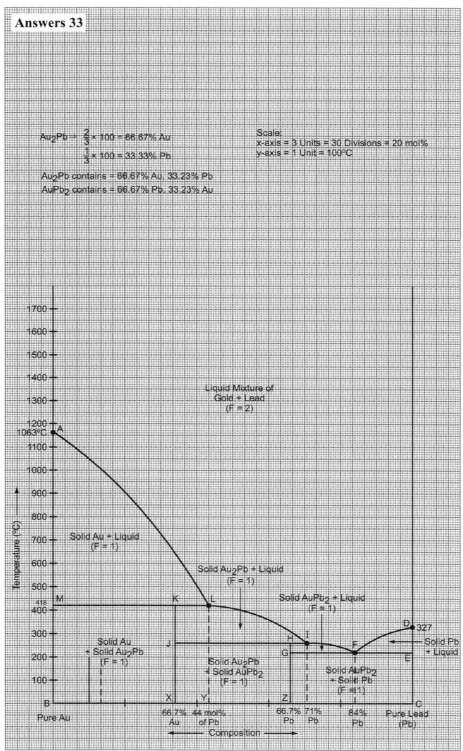

$Au_2Pb \rightarrow \dfrac{2}{3} \times 100 = 66.67\%$ Au

$\dfrac{1}{3} \times 100 = 33.33\%$ Pb

Au₂Pb contains = 66.67% Au, 33.23% Pb
AuPb₂ contains = 66.67% Pb, 33.23% Au

Scale:
x-axis = 3 Units = 30 Divisions = 20 mol%
y-axis = 1 Unit = 100°C

1700
1600
1500
1400
1300
1200
1063°C A
1100
1000
900
800
700 Solid Au + Liquid
     (F = 1)
600
500 M   K   L
418
400
300
     Solid Au   J
     + Solid Au₂Pb
200 (F = 1)
100
     B   X   Y   Z
     Pure Au   66.7% 44 mol%   66.7% 71%   84%   Pure Lead
                Au   of Pb   Pb   Pb   Pb   (Pb)

Liquid Mixture of
Gold + Lead
(F = 2)

Solid Au₂Pb + Liquid
(F = 1)

Solid AuPb₂ + Liquid
(F = 1)

Solid AuPb₂
+ Liquid
(F = 1)

D 327
Solid Pb
+ Liquid

Solid Au₂Pb
+ Solid AuPb₂
(F = 1)

Solid AuPb₂
+ Solid Pb
(F = 1)

Temperature (°C)

Composition

Question 33: Construct a phase diagram of gold-lead system from the following data: M.P. of Au = 1163°C, M.P. of lead = 327°C. Compound Au_2Pb decomposes at 418°C to give a liquid at 44 mol% of Pb. $AuPb_2$ has a peritectic at 254°C and 71% of Pb. Eutectic is at 215°C and 84% of Pb.

Explanation:

Curve *AL*: Freezing point curve of Au. Solid Au $\rightleftharpoons$ liquid mixture, $F = 1$ (monovariant).

Curve *DF*: Freezing point curve of lead, $F = 1$, solid Pb $\rightleftharpoons$ liquid.

Curve *FI*: Freezing point curve of $AuPb_2$, $F = 1$, solid Au_2Pb $\rightleftharpoons$ liquid.

Curve *LI*: Freezing point curve of Au_2Pb, $F = 1$, solid Au_2Pb $\rightleftharpoons$ liquid.

Point *F*: Eutectic point of $AuPb_2$ and Au, $F = 0$, solid $AuPb_2$ $\rightleftharpoons$ liquid $\rightleftharpoons$ solid Pb.

Point *I*: Peritectic point where $AuPb_2$ starts decomposing into Au_2Pb. Au_2Pb $\rightleftharpoons$ $AuPb_2$ $\rightleftharpoons$ liquid, $F = 0$.

Point *L*: Peritectic point, $F = 0$, solid Au_2Pb $\rightleftharpoons$ solid Au $\rightleftharpoons$ liquid.

Line *GE*: Eutectic line where solidification of Pb and $AuPb_2$ takes place, $F = 0$, solid $AuPb_2$ $\rightleftharpoons$ solid Pb $\rightleftharpoons$ liquid.

Area *ALKM*: Any point in this area represents solid Au $\rightleftharpoons$ liquid, $F = 1$.

Area *DEF*: Any point in this area represents solid Pb $\rightleftharpoons$ liquid, $F = 1$.

Area *HIFG*: Any point in this area represents solid $AuPb_2$ $\rightleftharpoons$ liquid, $F = 1$.

Area *KLIJ*: Any point in this area represent Au_2Pb $\rightleftharpoons$ liquid, $F = 1$.

Area below line *MK*: Solid Au $\rightleftharpoons$ solid Au_2Pb, $F = 1$.

Area below line *JH*: Solid Au_2Pb $\rightleftharpoons$ solid $AuPb_2$, $F = 1$.

Area below *GEF*: Any point in this area represents solid $AuPb_2$ $\rightleftharpoons$ solid Pb, $F = 1$.

Area above *ALIFD*: Liquid mixture of gold and lead, $F = 2$.

Answers 34 A_2B contains = 66.67% of $A\left(\frac{2}{3}\right)$, 33.33% of $B\left(\frac{1}{3}\right)$

AB_2 contains = 33.33% of A, 66.67 % of B

$$AB_2 \rightarrow A = \frac{1}{3} \times 100 = 33.33\%$$

$$B = \frac{2}{3} \times 100 = 66.67\%$$

$$A_2B \rightarrow A = \frac{2}{3} \times 100 = 66.67\%$$

$$B = \frac{1}{3} \times 100 = 33.33\%$$

Scale
x-axis = 3 Units = 30 Divisions = 20 mol%,
y-axis = 1 Unit = 100°C

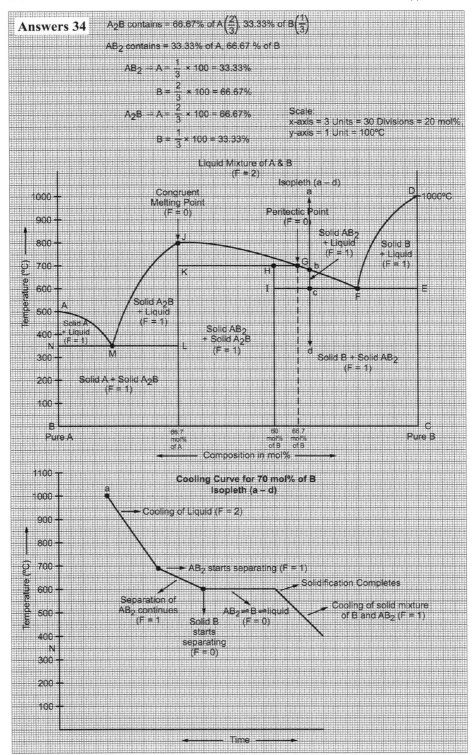

Question 34: A and B form two solid compounds A_2B and AB_2. Compound A_2B melts at 800°C to give a liquid having same composition as that of A_2B. The second compound AB_2 melts with B composition at 700°C to give solid B at 60 mol % of B. M.P. of $A = 500$°C, $B = 1000$°C. Sketch the phase diagram and draw the cooling curve for mixture of 70 mol% of B.

Explanation:

Curve AM: Freezing curve of A, solid $A \rightleftharpoons$ liquid, $F = 1$.

Curve DF: Freezing curve of B, solid $B \rightleftharpoons$ liquid, $F = 1$.

Curve GF: Freezing curve of AB_2, solid $AB_2 \rightleftharpoons$ liquid, $F = 1$.

Curve MJG: Freezing curve of A_2B, solid $A_2B \rightleftharpoons$ liquid, $F = 1$.

Line IE: Eutectic line, where solid $AB_2 \rightleftharpoons$ solid $B \rightleftharpoons$ liquid, $F = 0$.

Line NL: Eutectic line, where solid $A \rightleftharpoons A_2B \rightleftharpoons$ liquid, $F = 0$.

Point G: Peritectic point, solid $AB_2 \rightleftharpoons$ solid $A_2B \rightleftharpoons$ liquid, $F = 0$.

Point F: Eutectic point, solid $AB_2 \rightleftharpoons$ solid $B \rightleftharpoons$ liquid, $F = 0$.

Point M: Eutectic point, solid $A_2B \rightleftharpoons$ solid $A \rightleftharpoons$ liquid, $F = 0$.

Point J: Congruent melting point of solid A_2B, solid $A_2B \rightleftharpoons$ liquid, $F = 0$.

Area under line NL: Any point in this area represents solid $A \rightleftharpoons$ solid A_2B, $F = 1$.

Area under line IE: Any point in this area represents solid $B \rightleftharpoons$ solid AB_2, $F = 1$.

Area under line KH: Any point in this area represents solid $AB_2 \rightleftharpoons A_2B$.

Area AMN: Any point in this area represents solid $A \rightleftharpoons$ liquid, $F = 1$.

Area MLJ: Any point in this area represents solid $A_2B \rightleftharpoons$ liquid, $F = 1$.

Area $JKHG$: Any point in this area represents solid $A_2B \rightleftharpoons$ liquid, $F = 1$.

Area $HGFI$: Any point in this area represents solid $AB_2 \rightleftharpoons$ liquid, $F = 1$.

Area FDE: Any point in this area represents solid $B \rightleftharpoons$ liquid, $F = 1$.

Area Above $AMJGFD$: Liquid mixture of A and B, $F = 2$.

Answers 35

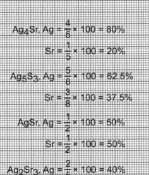

$$Ag_4Sr, Ag = \frac{4}{5} \times 100 = 80\%$$

$$Sr = \frac{1}{5} \times 100 = 20\%$$

$$Ag_5S_3, Ag = \frac{5}{8} \times 100 = 62.5\%$$

$$Sr = \frac{3}{8} \times 100 = 37.5\%$$

$$AgSr, Ag = \frac{1}{2} \times 100 = 50\%$$

$$Sr = \frac{1}{2} \times 100 = 50\%$$

$$Ag_2Sr_3, Ag = \frac{2}{5} \times 100 = 40\%$$

$$Sr = \frac{3}{5} \times 100 = 60\%$$

Scale:
x-axis = 3 Units = 30 Divisions = 20 mol%
y-axis = 1 Unit = 10 Divisions = 100°C

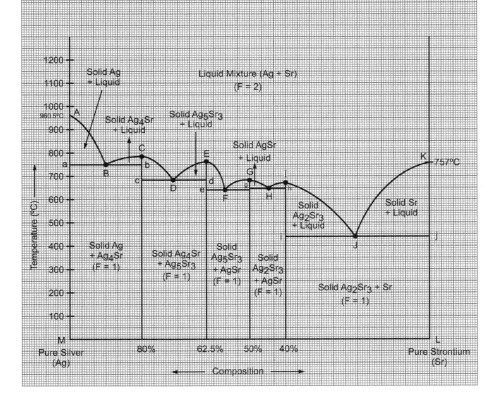

Question 35: Construct the phase diagram of silver strontium system.

M.P. of Ag –960.5°C

M.P. of Ag_4Sr –781°C

M.P. of Ag_5Sr_3 –760°C

M.P. of AgSr –680°C

M.P. of Ag_2Sr_3 –665°C

M.P. of Sr –757°C

750°C Eu_1,

693°C Eu_2

638°C Eu_3

645°C Eu_4

436°C Eu_5

Explanation:

Curve AB: Freezing point curve of Ag, solid Ag $\rightleftharpoons$ liquid, $F = 1$.

Curve JK: Freezing point curve of Sr, solid Sr $\rightleftharpoons$ liquid, $F = 1$.

Curve BCD: Freezing point curve of Ag_4Sr, solid Ag_4Sr $\rightleftharpoons$ liquid, $F = 1$.

Curve DEF: Freezing point curve of Ag_5Sr_3, solid Ag_5Sr_3 $\rightleftharpoons$ liquid, $F = 1$.

Curve FGH: Freezing point curve of AgSr, solid AgSr $\rightleftharpoons$ liquid, $F = 1$.

Curve HIJ: Freezing point curve of Ag_2Sr_3, solid Ag_2Sr_3 $\rightleftharpoons$ Liquid, $F = 1$.

Point B: Eutectic point, solid Ag $\rightleftharpoons$ solid Ag_4Sr $\rightleftharpoons$ liquid, $F = 0$.

Point D: Eutectic point, solid Ag_4Sr $\rightleftharpoons$ solid Ag_5Sr_3 $\rightleftharpoons$ liquid, $F = 0$.

Point F: Eutectic point, solid Ag_5Sr_3 $\rightleftharpoons$ solid AgSr $\rightleftharpoons$ liquid, $F = 0$.

Point H: Eutectic point, solid AgSr $\rightleftharpoons$ solid Ag_2Sr_3 $\rightleftharpoons$ liquid, $F = 0$.

Point J: Eutectic point, solid Ag_2Sr_3 $\rightleftharpoons$ solid Sr $\rightleftharpoons$ liquid, $F = 0$.

Point C: Congruent melting point of Ag_4Sr, solid Ag_4Sr $\rightleftharpoons$ liquid, $F = 0$.

Point E: Congruent melting point of Ag_5Sr_3, solid Ag_4Sr_3 $\rightleftharpoons$ melt Ag_4Sr_3, $F = 0$.

Point G: Congruent melting point of AgSr, solid AgSr $\rightleftharpoons$ liquid, $F = 0$.

Point I: Congruent melting point of Ag_2Sr_3, solid Ag_2Sr_3 $\rightleftharpoons$ liquid, $F = 0$.

Answers 36

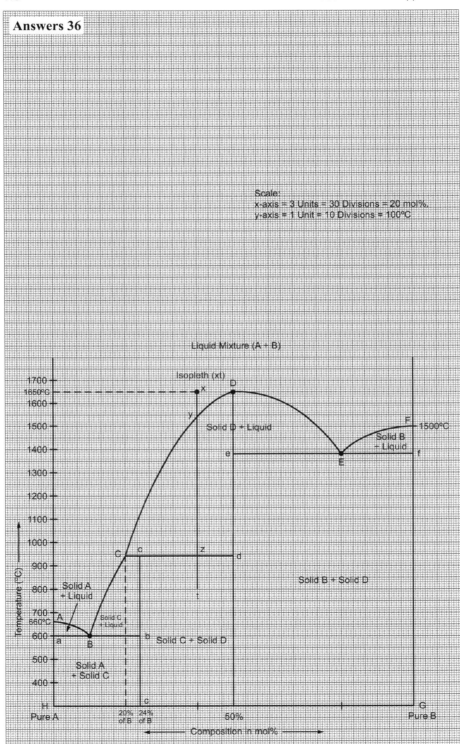

Scale:
x-axis = 3 Units = 30 Divisions = 20 mol%.
y-axis = 1 Unit = 10 Divisions = 100°C

Liquid Mixture (A + B)

Isopleth (xt)

1700
1650°C   x   D
1600
1500   Solid D + Liquid   F   1500°C
         y   Solid B
1400   e   + Liquid   f
                                              E
1300
1200
1100
1000
 900   C   c   z   d   Solid B + Solid D
 800   Solid A
        + Liquid
 700
 660°C A   Solid C
 600   a   B   + Liquid   b   Solid C + Solid D
 500   Solid A
             + Solid C
 400
        H   G
        Pure A   20%   24%   50%   Pure B
                  of B   of B
                        Composition in mol%

Question 36: The following information is known about the two-component A-B system. M.P. of pure $A = 660°C$, M.P. of pure $B = 1500°C$. Compound C having 24 mol% of B decomposes at 950°C to give D (50 mol% of B) and liquid containing 20 mol% of B. Compound D melts congruently at 1650°C. Eu_1 at 600°C, corresponding to composition of 10 mol% of B. $Eu_2 = 1380°C$ for 80 mol% of B. Draw the phase diagram. A melt containing 40 mol% of B is cooled from 1650°C, indicate the various phase changes which take place.

Explanation:

Curve AB: Freezing point curve of solid A, $F = 1$.

Curve EF: Freezing point curve of solid B, $F = 1$.

Curve CDE: Freezing point curve of solid D, $F = 1$.

Point B: Eutectic point, solid $A \rightleftharpoons$ solid $C \rightleftharpoons$ liquid, $F = 0$.

Point E: Eutectic point, solid $B \rightleftharpoons$ solid $D \rightleftharpoons$ liquid, $F = 0$.

Point C: Peritectic point, solid $C \rightleftharpoons$ solid $D \rightleftharpoons$ liquid, $F = 0$.

Point D: Congruent melting point, solid $D \rightleftharpoons$ melt ; $F = 0$.

Line ab: Eutectic line, $F = 0$ (Any point on this line represents solid A and solid C in equilibrium with liquid).

Line Cd: Transition of solid $C \rightleftharpoons$ solid D.

Line ef: Eutectic line, $F = 0$ (Any point on this line represents solid B and solid D in equilibrium with liquid.

Line cc': Solid C.

Area under ab: Any point in this area represents solid $A \rightleftharpoons$ solid C, $F = 1$.

Area under ef: Solid $B \rightleftharpoons$ solid D, $F = 1$.

Area under Cd: Any point in this area represents solid $C \rightleftharpoons$ solid D, $F = 1$.

Area above $ABCDEF$: Liquid mixture, $F = 2$.

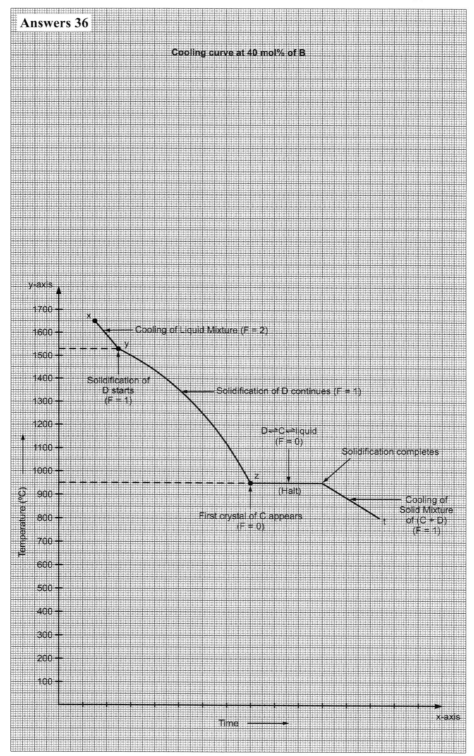

Answers 36

Cooling curve at 40 mol% of B

y-axis

Temperature (°C)

1700
1600 — x — Cooling of Liquid Mixture (F = 2)
1500 — y
1400
Solidification of
D starts — Solidification of D continues (F = 1)
1300 (F = 1)
1200
D⇌C+liquid
1100 (F = 0)
1000
Solidification completes
950 z
900 (Halt)
800 First crystal of C appears — Cooling of
(F = 0) t Solid Mixture
700 of (C + D)
600 (F = 1)
500
400
300
200
100

Time ⟶   x-axis

Explanation of cooling pattern (Line $x - t$)

- From x to y, temperature falls and cooling of liquid mixture of A and B takes place and system is bivariant.

- At y, solidification of D starts and it continues upto point z.

- At z, first crystal of C appears and the solidification continues for a certain period of time. Hence a halt occurs as there is no fall of temperature, $F = 0$.

- After completion of solidification till t, cooling of solid mixture takes place. Hence system here is monovariant.

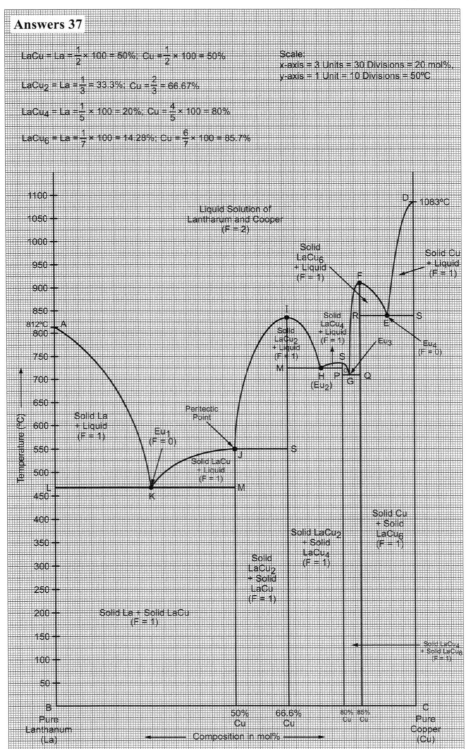

Answers 37

$LaCu = La = \frac{1}{2} \times 100 = 50\%; \ Cu = \frac{1}{2} \times 100 = 50\%$

$LaCu_2 = La = \frac{1}{3} = 33.3\%; \ Cu = \frac{2}{3} = 66.67\%$

$LaCu_4 = La = \frac{1}{5} \times 100 = 20\%; \ Cu = \frac{4}{5} \times 100 = 80\%$

$LaCu_6 = La = \frac{1}{7} \times 100 = 14.28\%; \ Cu = \frac{6}{7} \times 100 = 85.7\%$

Scale:
x-axis = 3 Units = 30 Divisions = 20 mol%,
y-axis = 1 Unit = 10 Divisions = 50°C

Liquid Solution of Lantharum and Cooper (F = 2)

Solid LaCu₆ + Liquid (F = 1)

Solid Cu + Liquid (F = 1)

1083°C

Solid LaCu₄ + Liquid (F = 1)

Solid LaCu₂ + Liquid (F = 1)

Solid La + Liquid (F = 1)

Peritectic Point

Eu₁ (F = 0)

Solid LaCu + Liquid (F = 1)

Eu₂

Eu₃

Eu₄ (F = 0)

Solid Cu + Solid LaCu₆ (F = 1)

Solid LaCu₂ + Solid LaCu₄ (F = 1)

Solid LaCu₂ + Solid LaCu (F = 1)

Solid La + Solid LaCu (F = 1)

Solid LaCu₄ + Solid LaCu₆ (F = 1)

Temperature (°C)

1100, 1050, 1000, 950, 900, 850, 812°C, 800, 750, 700, 650, 600, 550, 500, 450, 400, 350, 300, 250, 200, 150, 100, 50

Pure Lanthanum (La)

50% Cu   66.6% Cu   80% Cu   85% Cu   Pure Copper (Cu)

Composition in mol%

Question 37: The following data were obtained for the Cu-La system:

M.P. of Cu = 1083°C, M.P. of La = 812°C, The compound $LaCu_6$ melts congruently at 913°C. The compound $LaCu_4$ melts congruently at 735°C. The compound $LaCu_2$ melts congruently at 834°C. The compound LaCu melts incongruently at 551°C. The following eutectic temperatures were observed:

- 840°C (Cu + $LaCu_6$) • 725°C ($LaCu_6$ + $LaCu_4$)
- 468°C (LaCu + La). Sketch the phase diagram and label each area.

Explanation:
Curve AK and JK are the freezing point curves of La and LaCu, where F=1
Curves JIH, HSG and GFE are the freezing point curves of $LaCu_2$, $LaCu_4$ and $LaCu_6$ respectively, **JIH:** $LaCu_2 \rightleftharpoons$ liquid; **HSG:** $LaCu_4 \rightleftharpoons$ liquid; **GEF:** $LaCu_6 \rightleftharpoons$ liquid. Therfore, F=1 in all three cases.
Curve DE: Freezing point curve of solid Cu. It represents equilibrium between solid Cu and liquid, $F = 1$.

Point K: Eutectic point, solid La $\rightleftharpoons$ solid LaCu $\rightleftharpoons$ liquid, $F = 0$.

Point H: Eutectic point, solid $LaCu_2 \rightleftharpoons$ solid $LaCu_4 \rightleftharpoons$ liquid, $F = 0$.

Point G: Eutectic point, solid $LaCu_4 \rightleftharpoons$ solid $LaCu_6 \rightleftharpoons$ liquid, $F = 0$.

Point E: Eutectic point, solid Cu $\rightleftharpoons$ solid $LaCu_6 \rightleftharpoons$ liquid, $F = 0$.

Point J: Peritectic point, solid LaCu $\rightleftharpoons$ solid $LaCu_2 \rightleftharpoons$ liquid, $F = 0$.

Point I, S and F: Congruent melting points of $LaCu_2$, $LaCu_4$ and $LaCu_6$, F=0 in all three cases.

Area AKL: Any point in this area represents solid La $\rightleftharpoons$ liquid, $F = 1$.

Area KJM: Any point in this area represents solid LaCu $\rightleftharpoons$ liquid, $F = 1$.

Area JIHMJ: Any point in this area represents solid $LaCu_2 \rightleftharpoons$ liquid, $F = 1$.

Area HSGOH: Any point in this area represents solid $LaCu_4 \rightleftharpoons$ liquid, $F = 1$.

Area GFERQ: Any point in this area represents solid $LaCu_6 \rightleftharpoons$ liquid, $F = 1$.

Area DSED: Any point in this area represents solid Cu $\rightleftharpoons$ liquid, $F = 1$.

Area below LM: Any point in this area represents solid La $\rightleftharpoons$ solid LaCu, $F = 1$.

Area below JS: Any point in this area represents solid $LaCu_2 \rightleftharpoons$ solid LaCu, $F = 1$.

Area below HO: Any point in this area represents solid $LaCu_2 \rightleftharpoons$ solid $LaCu_4$, $F = 1$.

Area below PQ: Any point in this area represents solid $LaCu_4 \rightleftharpoons$ solid $LaCu_6$, $F = 1$.

Area below RS: Any point in this area represents solid Cu $\rightleftharpoons$ solid $LaCu_6$, $F = 1$.

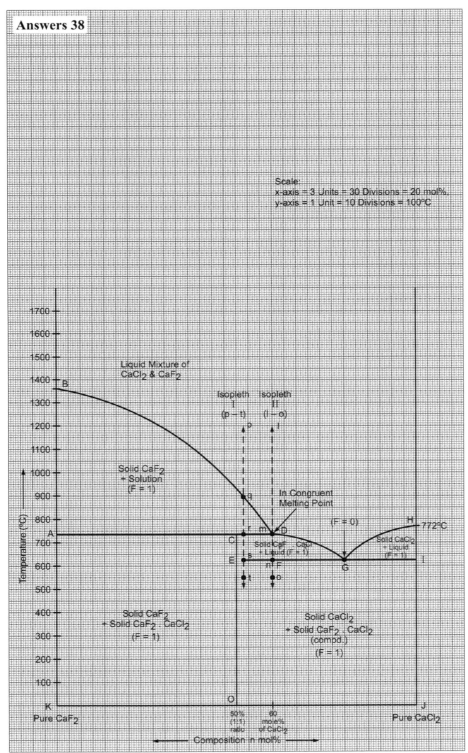

Answers 38

Scale:
x-axis = 3 Units = 30 Divisions = 20 mol%,
y-axis = 1 Unit = 10 Divisions = 100°C

Liquid Mixture of
CaCl$_2$ & CaF$_2$

Isopleth   Isopleth
I   II
(p − t)   (l − o)

Solid CaF$_2$
+ Solution
(F = 1)

In Congruent
Melting Point

(F = 0)   H 772°C

Solid CaCl$_2$
+ Liquid
(F = 1)

Solid CaF$_2$ CaCl
+ Liquid (F = 1)

Solid CaF$_2$
+ Solid CaF$_2$. CaCl$_2$
(F = 1)

Solid CaCl$_2$
+ Solid CaF$_2$. CaCl$_2$
(compd.)
(F = 1)

Temperature (°C)

1700
1600
1500
1400 B
1300
1200
1100
1000
900
800 A
700 C
600 E
500
400
300
200
100
K

Pure CaF$_2$

50%
(1:1)
ratio

60
mole%
of CaCl$_2$

Pure CaCl$_2$

Composition in mol%

Question 38: The system CaF_2 (M.P. 1360°C) and $CaCl_2$ (M.P. 772°C) shows an incongruent behaviour, forming a compound at 1 : 1 ratio. The compound melts at 737°C giving a liquid containing 60 mol% of $CaCl_2$. The eutectic point is at 625°C with the eutectic composition of 80 mol% of $CaCl_2$. Draw the phase diagram and label it. Draw cooling curves for 52% and 60% by mol of $CaCl_2$.

Explanation:

Point D: Peritectic point, involving solid CaF_2, solid $CaF_2 \cdot CaCl_2$ and solution, $F = 0$.

Point G: Eutectic point, solid $CaCl_2 \rightleftharpoons CaF_2.CaCl_2 \rightleftharpoons$ liquid, $F = 0$.

Area BDA: Solid $CaF_2 \rightleftharpoons$ solution, $F = 1$.

Area CDGE: Solid $CaF_2.CaCl_2 \rightleftharpoons$ liquid, $F = 1$.

Area HIG: Solid $CaCl_2 \rightleftharpoons$ liquid, $F = 1$.

Area below AC: Solid $CaF_2 \rightleftharpoons CaF_2 \cdot CaCl_2$ $F = 1$.

Area below EI: Solid $CaCl_2 \rightleftharpoons CaF_2 \cdot CaCl_2$ $F = 1$.

Area above BDGH: Liquid mixture of $CaCl_2$ and CaF_2, $F = 2$.

Line EI: Eutectic line, representing equilibrium between solid $CaCl_2$, and solid $CaF_2.CaCl_2$ in equilibrium with liquid, $F = 0$.

Line CEO: Solid compound $CaF_2.CaCl_2$.

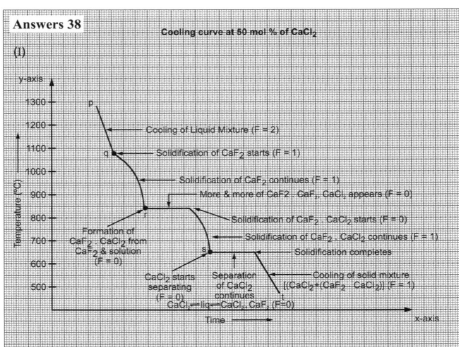

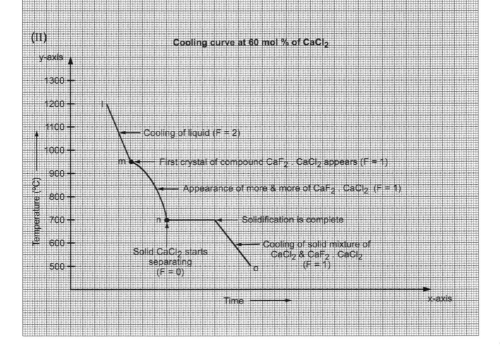

Explanation of cooling Patterns:

Isopleth I $(p - t)$

- From p to q cooling of liquid mixture takes place. At q, solidification of CaF_2 starts and continues till r, where first crystal of compound $CaF_2.CaCl_2$ appears. Then a halt comes and solidification of compound takes place till s.

- Now from s, solidification of $CaCl_2$ starts and again a halt appears without rise of temperature.

- Point s represents the eutectic line where $F = 0$.

- Fall of temperature is again noticed when solidification is complete and cooling of solid mixture takes place till point t.

Isopleth II $(l - o)$

- From l to m cooling of liquid mixture takes place and $F = 2$.

- At m, compound starts appearing and this point is called peritectic point. Till n solidification of compound continues.

- Now at n, solidification of $CaCl_2$ starts and continues till complete solidification.

- Cooling of solid mixture of $CaCl_2$ and $CaF_2.CaCl_2$ takes place upto point o.

Answers 39

$$A_2B_3 = A = \frac{2}{5} \times 100 = 40\%$$

$$B = \frac{3}{5} \times 100 = 60\%$$

$$AB_3 = A = \frac{1}{4} \times 100 = 25\%$$

$$B = \frac{3}{4} \times 100 = 75\%$$

Scale:
x-axis = 3 Units = 30 Divisions = 20 mol%,
y-axis = 1 Unit = 10 Divisions = 100°C

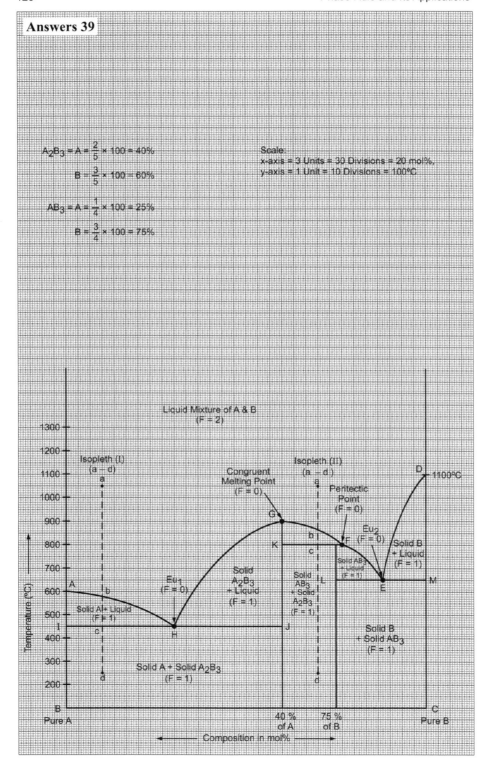

Question 39: Metals A and B form compounds AB_3 and A_2B_3. Solids A, B, AB_3 and A_2B_3 are in miscible in each other as solids but are completely miscible as liquids. Melting points of A and B are 600°C and 1100°C respectively. Compound A_2B_3 melts congruently at 900°C and gives a simple eutectic with A at 450°C. Compound AB_3 decomposes at 800°C to give another compound and the melt. This gives another eutectic with B at 650°C. Draw the simplest phase diagram with these facts and label. Draw cooling curve for melts of 90 mol% A and 30 mol% A.

Explanation:

Curve AH: Freezing point curve of solid A. Along this curve solid A in equilibrium with liquid, $F = 1$.

Curve DE: Freezing point curve of solid B, solid $B \rightleftharpoons$ liquid, $F = 1$.

Curve FE: Freezing point curve of solid AB_3. Along this curve solid $AB_3 \rightleftharpoons$ liquid and $F = 1$.

Curve HGF: Freezing point curve of solid A_2B_3 representing equilibrium between A_2B_3 and liquid, $F = 1$.

Point H: Eutectic point, involving solid $A \rightleftharpoons$ solid $A_2B_3 \rightleftharpoons$ liquid, $F = 0$.

Point E: Eutectic point, solid $B \rightleftharpoons AB_3 \rightleftharpoons$ liquid, $F = 0$.

Point F: Peritectic point involving solid AB_3, solid A_2B_3 and liquid, $F = 0$.

Point G: Congruent melting point of A_2B_3, solid $A_2B_3 \rightleftharpoons$ liquid, $F = 0$.

Line IJ: Eutectic line, a system on this line represents solid A $\rightleftharpoons$ solid $A_2B_3 \rightleftharpoons$ liquid, $F = 0$.

Line LM: Eutectic line, a system on this line represents solid $AB_3 \rightleftharpoons$ solid $B \rightleftharpoons$ liquid, $F = 0$.

Area AHI: Solid A $\rightleftharpoons$ liquid, $F = 1$.

Area $HGFKJ$: Solid $A_2B_3 \rightleftharpoons$ liquid, $F = 1$.

Area FLE: Solid $AB_3 \rightleftharpoons$ liquid, $F = 1$

Area DME: Solid $B \rightleftharpoons$ liquid, $F = 1$.

Area below IJ: Solid $A \rightleftharpoons$ solid A_2B_3, $F = 1$.

Area below KF: Solid $AB_3 \rightleftharpoons$ solid A_2B_3, $F = 1$.

Area below LM: Solid $B \rightleftharpoons$ solid AB_3, $F = 1$.

Area above $AHGFED$: Liquid mixture of A and B, $F = 2$.

Answers 39

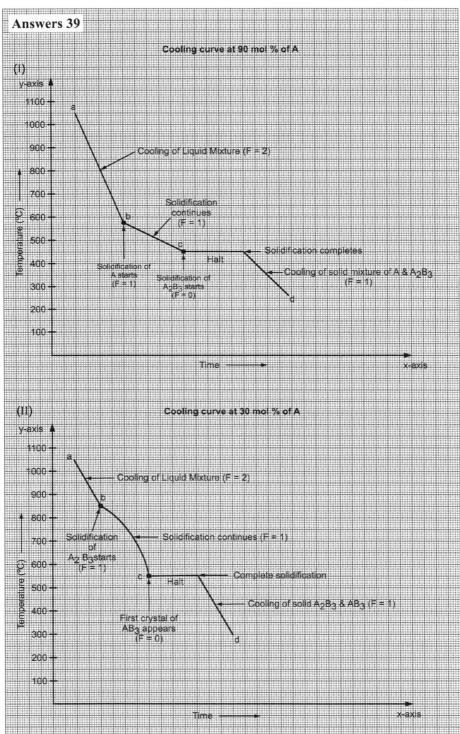

(I)

Cooling curve at 90 mol % of A

y-axis

Cooling of Liquid Mixture (F = 2)

Solidification continues (F = 1)

Solidification completes

Halt

Solidification of A starts (F = 1)

Solidification of A_2B_3 starts (F = 0)

Cooling of solid mixture of A & A_2B_3 (F = 1)

Temperature (°C)

Time

x-axis

(II)

Cooling curve at 30 mol % of A

y-axis

Cooling of Liquid Mixture (F = 2)

Solidification of A_2B_3 starts (F = 1)

Solidification continues (F = 1)

Complete solidification

Halt

Cooling of solid A_2B_3 & AB_3 (F = 1)

First crystal of AB_3 appears (F = 0)

Temperature (°C)

Time

x-axis

Explanation of cooling patterns:

Isopleth I $(a - d)$

- From a to b, cooling of liquid mixture takes place.

- At b, solidification of A starts and continues till c.

- At c, solidification of A_2B_3 starts which continues till halt line where $F = 0$.

- From here, cooling of solid mixture of A and A_2B_3 takes place upto d and system here is monovariant.

Isopleth II $(a' - d')$

- From a' to b', cooling of liquid mixture takes place.

- From b', solidification of A_2B_3 starts and continues till c'. System here is monovariant.

- From c', solidification of AB_3 starts and a halt comes without any rise of temperature. System is invariant.

- After the halt, complete solidification takes place and now cooling of solid mixture occurs and system here is monovariant.

Answers 40

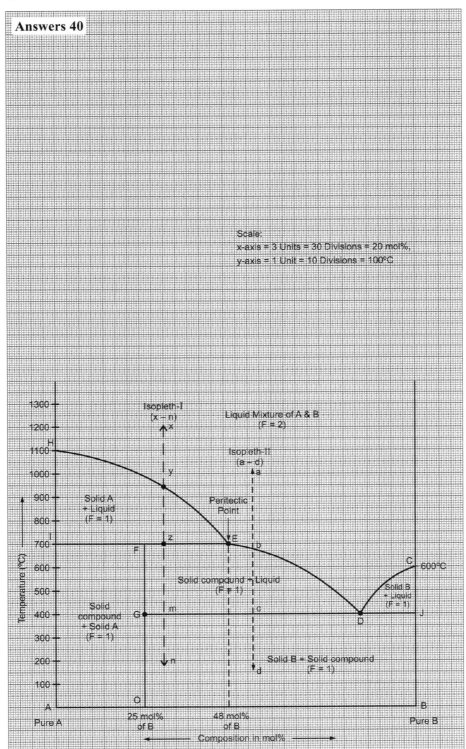

Scale:
x-axis = 3 Units = 30 Divisions = 20 mol%,
y-axis = 1 Unit = 10 Divisions = 100°C

Question 40: A system of A ($1100°C$) and B ($600°C$) shows incongruent behaviour. It has a peritectic temperature at $700°C$ where a solid containing 25 mol% of B melts giving a liquid of 48 mol% of B. The eutectic temperature is $400°C$ and the corresponding composition is 85 mol% of B. Plot the phase diagram. Draw cooling curves for 30 mol% of B and 55 mol% of B.

Explanation:

Curve *HE*: Freezing point curve of solid A, representing equilibrium between solid A and liquid and hence the system here is monovariant.

Curve *CD*: Freezing point curve of solid B involving solid B and liquid in equilibrium, $F = 1$.

Curve *DE*: Freezing point curve of solid compound. Along this curve, solid compound and liquid are in equilibrium and $F = 1$.

Point *D*: Eutectic point, solid $B \rightleftharpoons$ compound $\rightleftharpoons$ liquid, $F = 0$.

Point *E*: Peritectic point, solid $A \rightleftharpoons$ solid compound $\rightleftharpoons$ liquid, $F = 0$.

Line *GJ*: Eutectic line, a system on this line represents solid B and solid compound in equilibrium with liquid, $F = 0$.

Line *FO*: Solid compound.

Area *HIE*: Solid $A \rightleftharpoons$ liquid, $F = 1$.

Area *CDJ*: Solid $B \rightleftharpoons$ liquid, $F = 1$.

Area *FEDG*: Solid compound $\rightleftharpoons$ liquid, $F = 1$.

Area below *GDJ*: Solid $B \rightleftharpoons$ solid compound, $F = 1$

Area below *IF*: Solid compound $\rightleftharpoons$ solid A, $F = 1$.

Area above *HEDL*: Liquid mixture of A and B, $F = 2$.

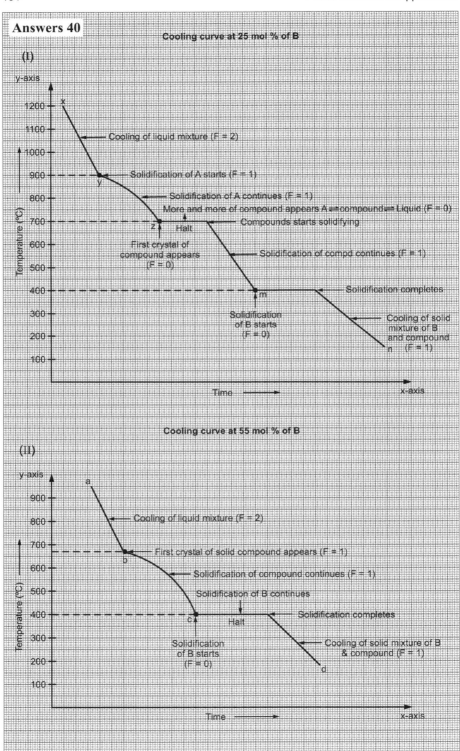

Answers 40

(I)

Cooling curve at 25 mol % of B

y-axis

x — Cooling of liquid mixture (F = 2)

Solidification of A starts (F = 1)

Solidification of A continues (F = 1)

More and more of compound appears A ⇌ compound ⇌ Liquid (F = 0)

Compounds starts solidifying

First crystal of compound appears (F = 0)

Halt

Solidification of compd continues (F = 1)

Solidification completes

Solidification of B starts (F = 0)

Cooling of solid mixture of B and compound (F = 1)

Temperature (°C)

Time ⟶   x-axis

(II)

Cooling curve at 55 mol % of B

y-axis

a — Cooling of liquid mixture (F = 2)

First crystal of solid compound appears (F = 1)

Solidification of compound continues (F = 1)

Solidification of B continues

Halt

Solidification completes

Solidification of B starts (F = 0)

Cooling of solid mixture of B & compound (F = 1)

Temperature (°C)

Time ⟶   x-axis

Explanation of cooling patterns:

Isopleth-I $(x - n)$

- From x to y, cooling of liquid mixture takes place.

- At y, solidification of A starts which continues till z and system is monovariant.

- At z, first crystal of compound appears, then a halt occurs and solidification of compound starts and continues till point m.

- At m, solidification of B starts and continues without rise of temperature (halt) and completes further followed by cooling of solid mixture upto n.

Isopleth-II $(a - d)$

- From a to b, cooling of liquid mixture takes place, and system here is monovariant.

- At b, solidification of compound starts which continues till c and system again is monovariant.

- At c, solidification of B starts and a halt occurs without rise of temperature and system here is invariant and represents eutectic line.

- After complete solidification, cooling of solid mixture takes place till point d and system here is monovariant.

Answers 41

$Mg_2Ni = Mg = \frac{2}{3} \times 100 = 66.6\%$

$Ni = \frac{1}{3} \times 100 = 33.3\%$

$MgNi_2 = Mg = \frac{1}{3} \times 100 = 33.3\%$

$Ni = \frac{2}{3} \times 100 = 66.6\%$

Scale:
x-axis = 3 Units = 30 Divisions = 20 mol%,
y-axis = 1 Unit = 10 Divisions = 100°C

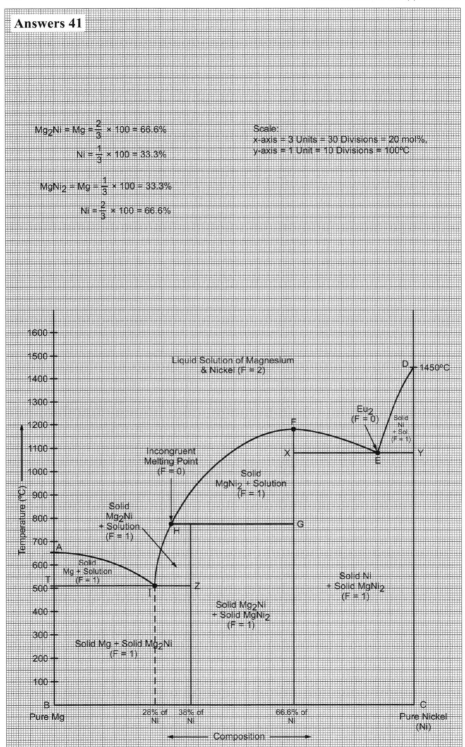

Question 41: The following data were obtained for Mg-Ni system: M.P. of Mg = 651°C M.P. of Ni = 1450°C Compound $MgNi_2$ melts congruently at 1180°C. Compound Mg_2Ni melts in congruently at 770°C giving a melt of 38 mass% of Ni. Eu_1 = 510°C at 28 mass% of Ni. Eu_2 = 1080°C at 88 mass% of Ni. No solid solution formation. Construct the phase diagram and label it.

Explanation:

Curve AI: Freezing point curve of solid Mg involving solid Mg and liquid in equilibrium, $F = 1$.

Curve DE: Freezing point curve of solid Ni involving solid Ni and liquid in equilibrium, $F = 1$.

Curve HI: Freezing point curve of compound Mg_2Ni. It represents equilibrium between Mg_2Ni and liquid, $F = 1$.

Curve HFE: Freezing point curve of $MgNi_2$. It represents equilibrium between $MgNi_2$ and liquid, $F = 1$.

Point I: Eutectic point, here solid Mg $\rightleftharpoons$ solid Mg_2Ni $\rightleftharpoons$ liquid, $F = 0$.

Point E: Eutectic point, solid Ni $\rightleftharpoons$ solid $MgNi_2$ $\rightleftharpoons$ liquid, $F = 0$.

Point F: Congruent melting point of solid $MgNi_2$, $F = 0$, solid $MgNi_2$ $\rightleftharpoons$ liquid.

Point H: Incongruent melting point, involving solid $MgNi_2$ and solid Mg_2Ni in equilibrium with liquid, $F = 0$.

Area AIT: Solid Mg $\rightleftharpoons$ solution, $F = 1$.

Area DYE: Solid Ni $\rightleftharpoons$ solution, $F = 1$.

Area HGXEF: Solid $MgNi_2$ $\rightleftharpoons$ solution system is monovariant.

Area IZH: Solid Mg_2Ni $\rightleftharpoons$ solution, system is monovariant.

Area below IZ: Solid Mg $\rightleftharpoons$ solid Mg_2Ni, system is monovariant.

Area below XEY: Solid Ni $\rightleftharpoons$ solid $MgNi_2$, system is monovariant.

Area below HG: Solid Mg_2Ni $\rightleftharpoons$ solid $MgNi_2$, system is monovariant.

Area above AIHFED: Liquid solution of Mg and Ni, system is bivariant.

Line TZ: Eutectic line, a system on this line represents solid Mg and Mg_2Ni in equilibrium with liquid, $F = 0$.

Line XY: Eutectic line, a system on this line represents solid Ni and solid $MgNi_2$ in equilibrium with liquid, $F = 0$.

Answers 42

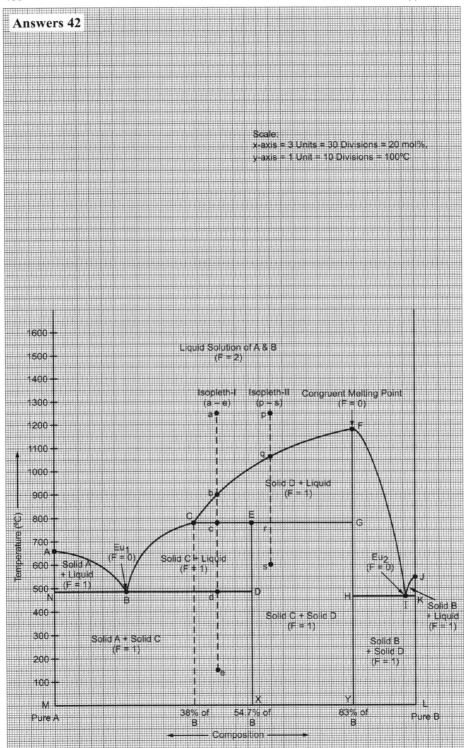

Scale:
x-axis = 3 Units = 30 Divisions = 20 mol%,
y-axis = 1 Unit = 10 Divisions = 100°C

Liquid Solution of A & B
(F = 2)

Isopleth-I   Isopleth-II   Congruent Melting Point
(a – e)   (p – s)   (F = 0)

Solid D + Liquid
(F = 1)

Eu₁
(F = 0)

Solid C + Liquid
(F = 1)

Solid A
+ Liquid
(F = 1)

Eu₂
(F = 0)

Solid B
+ Liquid
(F = 1)

Solid C + Solid D
(F = 1)

Solid A + Solid C
(F = 1)

Solid B
+ Solid D
(F = 1)

Pure A   38% of   54.7% of   83% of   Pure B
                      B   B   B

Composition

Temperature (°C)

Question 42: The following information is known about the two-component A-B system: Atomic mass of $A = 24.39 g/mol$, M.P. of $A = 651°C$ Atomic mass of $B = 59.7 g/mol$, M.P. of $B = 550°C$. Two compounds C and D are formed. C contains 54.7% by mass of B and melts incongruently at 770°C yielding D and a liquid containing 38% by mass of B. D melts sharply at 1180°C yielding a liquid containing 83% by mass of B. Solids are immiscible in the solid state and completely miscible in liquid phase. Construct a phase diagram and label it. Draw cooling curves for 45% and 60% by mass of B.

Explanation:

Curve AB: Freezing point curve of A, solid $A \rightleftharpoons$ liquid $F = 1$.

Curve JI: Freezing point curve of B, solid $B \rightleftharpoons$ liquid, $F = 1$.

Curve BC: Freezing point curve of C, solid $C \rightleftharpoons$ liquid, $F = 1$.

Curve CFI: Freezing point curve of D, solid $D \rightleftharpoons$ liquid, $F = 1$.

Point B: Eutectic point, solid $A \rightleftharpoons$ solid $C \rightleftharpoons$ liquid, $F = 0$.

Point I: Eutectic point, solid $B \rightleftharpoons$ solid $D \rightleftharpoons$ liquid, $F = 0$.

Point C: Peritectic point, involving solid C and solid D in equilibrium with liquid, $F = 0$.

Point F: Congruent melting point, solid $D \rightleftharpoons$ liquid, $F = 0$.

Line EDX: Compound C.

Line ND: Eutectic line, any point on this line represents solid A and solid C in equilibrium with liquid, $F = 0$.

Answers 42

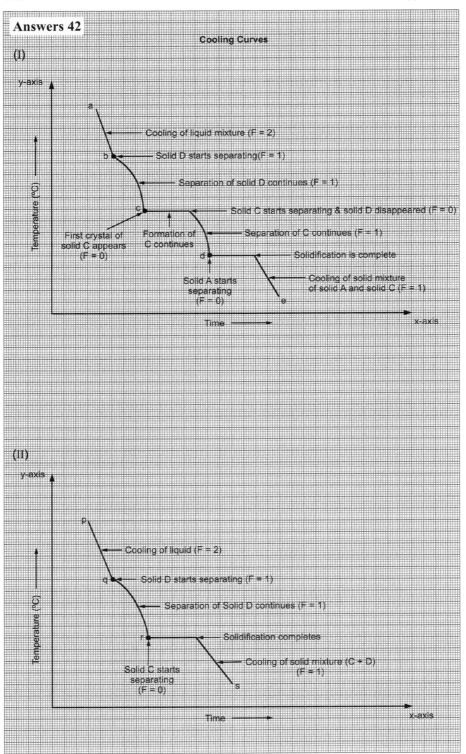

Cooling Curves

(I)

y-axis

Temperature (°C)

a

Cooling of liquid mixture (F = 2)

b — Solid D starts separating(F = 1)

Separation of solid D continues (F = 1)

c — Solid C starts separating & solid D disappeared (F = 0)

First crystal of solid C appears (F = 0)

Formation of C continues — Separation of C continues (F = 1)

d — Solidification is complete

Solid A starts separating (F = 0)

Cooling of solid mixture of solid A and solid C (F = 1)

e

Time ———→   x-axis

(II)

y-axis

Temperature (°C)

p

Cooling of liquid (F = 2)

q — Solid D starts separating (F = 1)

Separation of Solid D continues (F = 1)

r — Solidification completes

Solid C starts separating (F = 0)

Cooling of solid mixture (C + D) (F = 1)

s

Time ———→   x-axis

Line *HK*: Eutectic line, any point on this line represents solid B and solid D in equilibrium with liquid, $F = 0$.

Area below *NBD*: Any point in this area represents solid $A \rightleftharpoons$ solid C, $F = 1$. Here cooling of solid A and solid C takes place.

Area below *HKL*: Any point in this area represent solid $B \rightleftharpoons$ solid D, $F = 1$. Here cooling of solid B and D takes place.

Area below *EG*: Any point in this area represent solid $C \rightleftharpoons$ solid D. $F = 1$. Here cooling of solid C and D takes place.

Area *ABN*: It represent solid $A \rightleftharpoons$ liquid, $F = 1$.

Area *BCED*: It represent solid $C \rightleftharpoons$ liquid, $F = 1$.

Answers 43

$CaAl_3 = Ca = \dfrac{1}{4} \times 100 = 25\%$

$Al = \dfrac{3}{4} \times 100 = 75\%$

$CaAl_2 = Ca = \dfrac{1}{3} \times 100 = 33.3\%$

$Al = \dfrac{2}{3} \times 100 = 66.6\%$

Scale:
x-axis = 1 Units = 10 Divisions = 10 mol%
y-axis = 1 Unit = 10 Division = 100°C

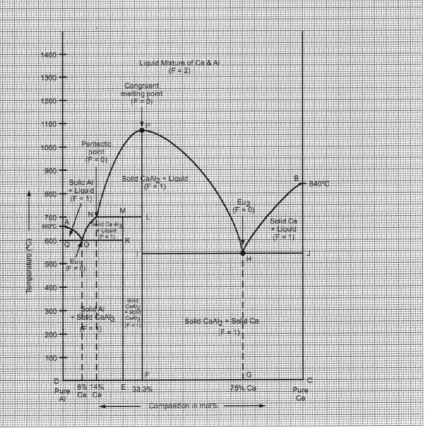

Question 43: (*a*) The phase diagram of Ca-Al system has following data: M.P. of Ca is 840°C, M.P. of Al is 660°C, congruent M.P. of $CaAl_2$ = 1063°C, incongruent M.P. of $CaAl_3$ = 700°C and composition is 14 mass % Ca. Eutectic point involving Ca and $CaAl_2$ is 540°C and has 75 mass % of Ca. Eutetic point involving Al and $CaAl_3$ is at 600°C and has 8 mass % of Ca.

(*b*) Draw phase diagram and from it predict the nature of the system for the following data:

(*i*) 80 mass % of Ca at 700°C   (*vi*) 25 mass % of Ca at 650°C

(*ii*) 60 mass % of Ca at 800°C   (*vii*) 25 mass % of Ca at 500°C

(*iii*) 37 mass % of Ca at 800°C   (*viii*) 10 mass % of Ca at 640°C

(*iv*) 37 mass % of Ca at 600°C   (*ix*) 5 mass % of Ca at 640°C

(*v*) 25 mass % of Ca at 750°C   (*x*) 5 mass % of Ca at 500°C

Explanation

(a):

Curve *AO*: Freezing point curve of Al, $F = 1$.

Curve *HB*: Freezing point curve of Ca, $F = 1$.

Curve *NPH*: Freezing point curve of $CaAl_2$, $F = 1$.

Curve *NO*: Freezing point curve of $CaAl_3$, $F = 1$.

Point *O*: Eutectic point, solid Al $\rightleftharpoons$ solid $CaAl_3$ $\rightleftharpoons$ liquid, $F = 0$.

Point *H*: Eutectic point, solid Ca $\rightleftharpoons$ solid $CaAl_2$ $\rightleftharpoons$ liquid, $F = 0$.

Point *N*: Peritectic point, solid $CaAl_3$ $\rightleftharpoons$ solid $CaAl_2$ $\rightleftharpoons$ liquid, $F = 0$.

Point *P*: Congruent melting point, liquid $\rightleftharpoons$ solid $CaAl_2$, $F = 0$.

Area below *QK*: Solid Al $\rightleftharpoons$ solid $CaAl_3$, $F = 1$.

Area below *ML*: Solid $CaAl_3$ $\rightleftharpoons$ solid $CaAl_2$, $F = 1$.

Area below *IJ*: Solid $CaAl_2$ $\rightleftharpoons$ solid Ca, $F = 1$.

Area *AOQ*: Solid Al $\rightleftharpoons$ liquid, $F = 1$.

Area *NMKO*: Solid $CaAl_3$ $\rightleftharpoons$ liquid, $F = 1$.

Area *NMLIHP*: Solid $CaAl_2$ $\rightleftharpoons$ liquid, $F = 1$.

Area *HJB*: Solid Ca $\rightleftharpoons$ liquid, $F = 1$.

Area above *AONPHB*: Liquid mixture of Ca and Al, $F = 2$.

(b):

(*i*) 80 mass % of Ca at 700°C: consists all liquid mixture of Ca and Al, with $F = 2$.

(*ii*) 60 mass % of Ca at 800°C: consists of liquid and solid $CaAl_2$, $F = 1$.

(*iii*) 37 mass% of Ca at 800°C: consists of liquid and solid $CaAl_2$, $F = 1$.

(*iv*) 37 mass % of Ca at 600°C: consists of liquid and solid $CaAl_2$, $F = 1$.

(*v*) 25 mass % of Ca at 750°C: consists of solid $CaAl_2$ and liquid, $F = 1$.

(*vi*) 25 mass % of Ca at 650°C: Solid $CaAl_2$ is in equilibrium with solid $CaAl_3$ which is in equilibrium with liquid, $F = 0$.

(*vii*) 25 mass % Ca at 500°C: Solid Al $\rightleftharpoons$ Solid $CaAl_2$ $\rightleftharpoons$ Solid $CaAl_3$, $F = 0$.

(*viii*) 10 mass % of Ca at 640°C: Solid $CaAl_3$ $\rightleftharpoons$ Liquid, $F = 1$.

(*ix*) 5 mass % of Ca at 640°C: Solid Al $\rightleftharpoons$ Liquid, $F = 1$.

(*x*) 5 mass % of Ca at 500°C: Solid Al $\rightleftharpoons$ Solid $CaAl_3$, $F = 1$.

EXPERIMENT ON PHASE DIAGRAM

Aim: To construct a phase diagram for a two component system (simple eutectic) by plotting cooling curves for mixture of different compositions (benzoic and cinnamic acid) using thermal analysis method.

Apparatus Required: Thermometer (360°C), boiling tubes, a bath of some colourless liquid (liquid paraffin or glycerine or any refined oil or any vanaspati ghee), beaker, an inert wire stirrer, an iron stand with clamp, tripod stand with wire gauze, a bunsen burner.

PROCEDURE

(*i*) Prepare the following mixtures of *A* and *B* components by weighing the requisite quantities in the boiling tube.

S.No.	Mass % of cinnamic acid (B)	Mass of benzoic acid A (g)	Mass of cinnamic acid B (g)
1	100	6	5
2	80	4	4
3	60	2	3
4	50	2.5	2.5
5	40	3	2
6	20	4	1
7	0 (100% of A)	5	0

(*ii*) In a clear dry boiling tube, take about 5 g of benzoic acid *i.e.*, 100% benzoic and (component A) and clamp it vertically with the help of a clamp stand. Insert the thermometer and the stirrer into the boiling tube using cork. (Use cotton if cork is not available)

(*iii*) Heat the contents of the boiling tube by placing it in oil bath slightly above the melting point of the pure component A. When all the crystals of the pure component *A* have disappeared and a homogeneous liquid phase is obtained, remove the boiling tube from the oil bath and wipe it clean with the help of either filter paper or tissue paper.

(*iv*) Stir the mixture slowly with the help of a stirrer and note down the temperature after every half minute.

(*v*) Note the temperature at which first crystals of the solid are formed which signifies the break point.

(*vi*) Allow the cooling to continue and find out the temperature or temperatures at which the liquid solid system shows temperature halts (which is shown by the constancy or arrest in temperature). This results in a complete arrest of the mixture. The lowest temperature halt will be its eutectic temperature. At this temperature, the composition of the solid phase which separates will have the same composition as that of the liquid phase. The temperature of the system will remain constant until the entire liquid phase has solidified. After this, the system will exhibit smooth cooling of the solid phase.

(*vii*) Repeat the above procedure for other compositions as given in the table.

(*viii*) Construct the cooling curves for different compositions by plotting a graph between the temperature (°C) on *Y*-axis and the time (sec) on *X*-axis.

(*ix*) From the cooling curves for different compositions, note down the break and halts and then plot a graph between temperature in °C (on *Y*-axis) and percentage composition (on *X*-axis) to get a phase diagram of the two component system.

The experiment has been performed under laboratory conditions and the following data was recorded.

OBSERVATION TABLE (a)

Time (min)	a Pure A 100% benzoic acid	b 20% cinnamic acid	c 40 % cinnamic acid	d 50% cinnamic acid	e 60% cinnamic acid	f 80% cinnamic acid	g 100% cinnamic acid
0.5	147	156	152	154	154	156	162
1.0	136	145	139	144	145	146	150
1.5	128	135	128	132	135	135	140
2.0	120	124	117	123	125	126	131
2.5	120	114	109	115	117	118	130
3.0	120	110	101	108	108	114	130
3.5	120	108	93	100	102	111	130
4.0	119	105	89	95	96	108	128
4.5	115	103	87	90	90	105	126
5.0	112	99	83	85	84	98	122
5.5	105	95	80	80	80	96	119
6.0	99	91	78	78	80	90	114
6.5		87	78	78	78	84	
7.0		80	78	78	78	78	
7.5		80	78	78	78	79	
8.0		78	78	78	78	78	
8.5		78	77	76	77	78	
9.0		73	75	71	75	78	
9.5		73	75	71	75	78	
10.0		71	74	68	74	78	
10.5		69	72	65	73	75	
11.0			79			70	
12.0						68	

OBSERVATION TABLE (b)

S.No.	Percentage composition B (cinnamic acid)	Temperature (°C)	
		Break	Halt
1	20%	114	78
2	40%	93	78
3	50%	94	78
4	60%	108	78
5	80%	103	78

RESULT

The cooling curves and the phase diagram for benzoic-cinnamic acid are drawn. The eutectic temperature of the given system is 78°C and the corresponding eutectic composition of the mixture is 54% of benzoic acid and 46% of cinnamic acid.

LITERATURE VALUE

Melting point of benzoic acid in °C = 121°C

Melting point of cinnamic acid in °C = 133°C

The eutectic temperature of benzoic-cinnamic system = 81°C

The eutectic composition is 56.5 mol % of benzoic acid and 43.5 mol % of cinnamic acid.

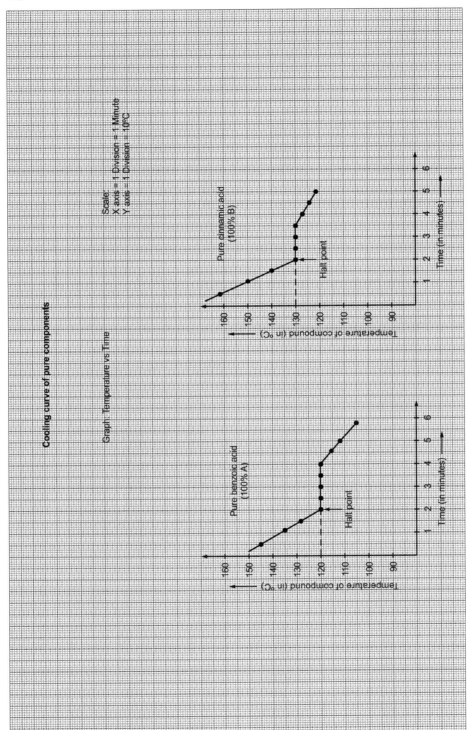

Cooling curve of pure components

Graph: Temperature vs Time

Scale:
X axis = 1 Division = 1 Minute
Y axis = 1 Division = 10°C

Pure cinnamic acid
(100% B)

Halt point

Time (in minutes)

Temperature of compound (in °C)

Pure benzoic acid
(100% A)

Halt point

Time (in minutes)

Temperature of compound (in °C)

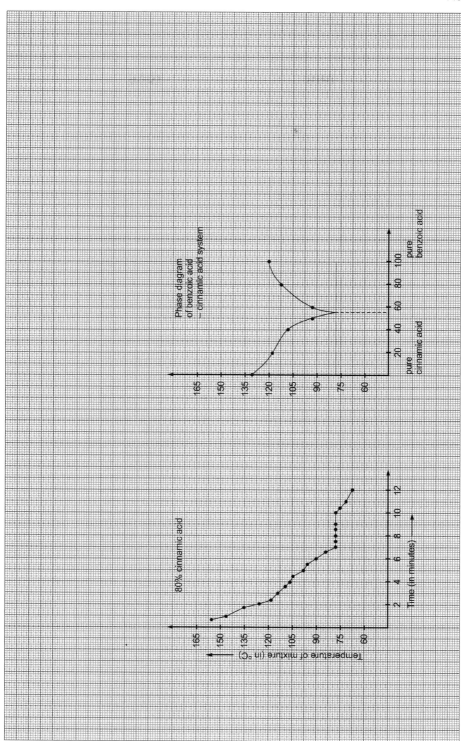

Index